Hey friend,

Thank you for picking up my book! I've poured everything I've learned about B2B onboarding from working with hundreds of companies into these pages, so it's incredibly fulfilling to know you're reading this.

Here's the deal: You can get all the templates, worksheets, and examples I mention throughout this book for FREE by scanning the QR code or visiting this link:

eurekabonus.com

All right, it's time to transform your onboarding into a bridge to customer success.

Ready?
Ramli "RJ" John

PRAISE FOR
EUREKA

"Before you can get users *HOOKED* on your product, you have to first ensure they start using it. *EUREKA* provides a detailed, practical, how-to guide to crafting the perfect onboarding experience."

—NIR EYAL,
bestselling author of Hooked and Indistractable

"For most SaaS companies, the real growth problem starts when people sign up and don't stick around. This is usually an onboarding problem, and it's often the biggest weakness for most businesses. *EUREKA* is a must-read for anyone embarking on the critical activity of improving their product onboarding."

—SEAN ELLIS,
founder of GrowthHackers.com
and author of Hacking Growth

"A refreshing, no-fluff, and incredibly practical guide to B2B SaaS onboarding that treats poor onboarding as what it truly is—a symptom of strategic misalignment. Ramli shares what feels like a trade secret. Packed with real-world insights, frameworks, and hard-earned wisdom, this book is a must-read. It is a gold mine, especially for sales-led and mid-market to enterprise-focused teams."

—VIVEK BALASUBRAMANIAN,
VP of Product Growth at Roofr and
former growth leader at Wave Apps and Shopify

"Ramli John is one of the top experts on user onboarding or product adoption. He has a unique perspective and framework to help companies turn users into loyal fans. If you'd like to learn more about product onboarding, Ramli is the go-to person. Get his book now!"

—**KATE SYUMA**,
founder of Growthmates and
former Head of Growth Design, Miro

"Ramli provides a step-by-step guide on how to build a proper onboarding strategy. Having worked at Hotjar, Sprout Social, Teamwork, and now Bitly, I appreciate the level of detail and accountability Ramli lays out for the entire organization. With this book, our teams now have a playbook that can be easily followed and absolutely nails every step in the process (in an easily digestible way). I can't wait to buy this book for my entire organization and couldn't recommend it more!"

—**TARA ROBERTSON**,
CMO at Bitly

"One of the most valuable things I've seen as a Customer Success leader is the impact great user onboarding can have on long-term customer success, from driving adoption to reducing the load on support while increasing retention and customer advocacy. This book is hands down one of the best roadmaps I've seen to successful B2B onboarding. Implementing the best practices and techniques in *EUREKA* is an investment in your customers, your business, and your future!"

—**DANNY VILLARREAL**,
VP of Customer Success at Toolio

"Ramli is one of very few I would work with to help SaaS improve their onboarding. He knows that it takes a customer-led approach to first address the core issues that hinder adoption and retention. Don't miss your chance to work with and learn from him."

—**GEORGIANA LAUDI**,
co-founder and author,
Forget the Funnel and CX Advisor

"If a company is taking off, Ramli studies its onboarding. Having done so for so many years, he has built up an industry-leading perspective on what does and doesn't work. I can't think of a better book to improve your product onboarding."

—**AAKASH GUPTA**,
CPO, Product Growth and
former VP of Product, Apollo.io

"Our Growth Team at Wistia spent a ton of time working on new user onboarding. We knew how important it was to our success, and we had a hunch that our approach could be more effective. We had a lot of passion, but in retrospect—we had no real plan or framework to guide our thinking. This is the book I wish our team had. *EUREKA* breaks down everything a product-led business needs to know about onboarding—from understanding your users, defining success criteria, driving engagement, and optimizing along the way. It's full of actionable takeaways that you can apply today for a big impact."

—**ANDREW CAPLAND**,
coach for growth teams, PLG Advisor, former
Growth at Wistia, Postscript

"I have picked up so many valuable insights from this book that I now use in my work as a Growth Designer. Ramli explains concepts thoroughly and shares examples that are super helpful. The templates from the book are really useful in my day-to-day work. If you want to improve your onboarding skills and learn from the best, I highly recommend *EUREKA*!"

—BRITNEY MACKEY,
Growth Product Designer, Zapier

"Retention is the foundation of sustainable growth, and it starts the moment someone perceives value in your product. This book provides you with a framework to optimize and build on that moment to raise the odds of people making your product an integral part of how they achieve their goals."

—ANUJ ADHIYA,
author of Growth Hacking for Dummies, Growth Mentor at First Round Fast Track, and VP Growth at Sophya

"Ramli hosted a workshop for our team. I have to say that I just saw his superpower! Ramli guided us through identifying roadblocks in our free trial experience and coming up with solutions to engage and convert our trial users. I highly recommend him if you need a solid strategy to onboard and convert your users."

—JACKSON NOEL,
co-founder and former CEO, Appcues

"Ramli is one of the top product-led growth experts. He gives insightful and actionable advice that can help you uncover hidden opportunities for growth and optimize your user onboarding experience. Ramli has a structured approach to creating engaging and effective onboarding flows. Whether you're a startup looking to establish a strong foundation or an established company aiming to revitalize your user experience, work with Ramli now. *EUREKA* can help you create an onboarding journey that drives growth and retention."

—**MARIO ARAUJO**,
Head of Product Growth, Penpot

"As any business thinks about word of mouth and referrals, one must think about the whole customer experience—not just the numbers side but the individual experience of each person and their motivations. What Ramli has done with this book is to explain a practical and relatable framework for onboarding that drives real business value. It's clear through all aspects of the book that this isn't just one person's job, but Ramli makes it easy to include multiple team members to drive a collaborative process."

—**JOSH HO**,
founder and CEO, ReferralRock

"Ramli is a seasoned growth practitioner, and it shows. *EUREKA* shines with stories from his learnings and practical knowledge that make some of the growth's biggest head-scratchers digestible and human. If you're looking to up-level your onboarding or invest in scaling your startup, this is a great read."

—**EMILY LONETTO**,
Sr. Director of Community
and Agencies, Webflow

"If there's anyone who knows the most about onboarding, it's Ramli. His EUREKA Framework helped me significantly improve WordPress's activation rate and free-to-paid conversion. If your company struggles with those and wants a clear direction and strategy, Ramli's your man. Get this book now!"

—**PHIL GAMACHE**,
founder, Humans of Martech
and former Director of Growth, WordPress.com

"Onboarding has been one of the most critical stages of driving growth for the companies I advise and for the teams I've led at fast-growing PLG startups like Oyster and Buffer. And Ramli is one of the very best in the onboarding business! He has a process and strategy in his book that is brilliant for turning signups into active, engaged users. I'm so impressed by his onboarding expertise, and I'm even more appreciative for the humble, thoughtful, curious approach he takes to his work. Any company would be very lucky to get the chance to partner with him!"

—**KEVAN LEE**,
co-founder, Bonfire and
former VP of Marketing, Buffer

"Ramli John is a true expert in user onboarding and product adoption. He doesn't just provide theoretical concepts; he offers a step-by-step guide that any team can implement immediately. And he makes it all easy to understand with really engaging examples. If you're serious about improving your product's onboarding and driving exponential growth, Ramli is the expert you need."

—**MEG GOWELL**,
Director of Growth Marketing, Typeform

"Ramli's extremely creative and has bold ideas to help any SaaS company turn its product into a growth engine. He's a natural educator and knows how to make an impact on other people's lives and businesses."

—WES BUSH,
Founder and CEO, ProductLed

"A must-read for any software organization. *EUREKA* takes the daunting task of user onboarding and turns it into an actionable step-by-step guide that enables Marketing, Product, and Engineering teams to work together. Our team got a chance to personally work with Ramli to help us reduce the number of steps in our welcome tour, and he gave us guidance on improving our entire onboarding flow. We paid thousands of dollars to work with Ramli, and now you can read his process for less than that."

—AMANDA NATIVIDAD,
VP of Marketing, SparkToro

"Ramli is one of the few people who actually GETS onboarding and its impact on growth. He takes the time to understand, is incredibly curious to learn more, and goes out of his way to share that wisdom."

—SHAREIL NARIMAN,
Director of Customer Experience, Arrows

"Ramli identified the bottlenecks in our onboarding and developed actionable strategies to streamline it. Thanks to him, we have a plan to reduce our customer onboarding time from weeks to days. This book will help you achieve the same results!"

—SUSAN FANCHER,
Director of Customer Success,
SE Healthcare

"Look. There are a lot of pretenders out there trying to tell you what to do. Ramli isn't one of them. I've known Ramli for many, many years now, and he is one of the most gifted, hard-working growth leaders I know. Some people focus on strategy, some people are obsessed with tactics. This book will give you both!"

—MARC THOMAS,
founder, Positive Human and
former Senior Growth
Marketing Manager, Podia

"I've been following Ramli for years and read his excellent book. I can confidently say that Ramli is a master of the onboarding craft! You'd be lucky to work with him."

—KEVIN INDIG,
Growth Advisor, Dropbox,
Riverside.fm, and Shopify

"Folks, Ramli is not just a thought leader—he's the real deal. Ramli can take yucky, confusing, and stress-inducing onboarding experiences and make them user-friendly and effective."

—LYLA ROZELLE,
Senior Product Manager, Quickbase and
former Senior Product Growth Manager, Litmus

"Ramli has seen it, done it, and wrote the authoritative book on product onboarding. If you're considering partnering with someone to help take your onboarding to the next level, Ramli has to be top of your list."

—BEN WILLIAMS,
founder, The Product-Led Geek
and Former VP of Product, Snyk

"After one session with Ramli, our team got practical insights and actionable steps to improve our onboarding process. His expertise in user onboarding and activation is noticeable and obvious. Ramli provides great suggestions to create engaging onboarding experiences that drive long-term user success. I'd recommend *EUREKA* to anyone looking to increase user activation or optimize their funnel."

—ANAND PATEL,
Director of Product Marketing,
Goldcast

"The insights I gained, especially regarding my company onboarding strategies, have been a game changer for me. *EUREKA* is not just informative but also engaging, offering a deep understanding of user behavior and needs."

—ALAN ARDUIN,
Principal Product Manager, Azion

EUREKA

The
product onboarding playbook
for High-Growth B2B companies

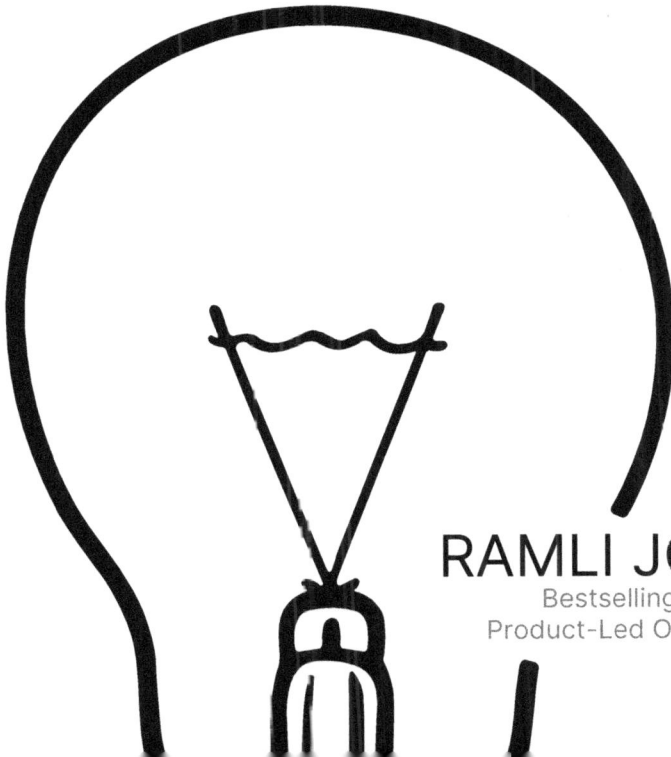

RAMLI JOHN
Bestselling author of
Product-Led Onboarding

Published by Damn Gravity Media LLC, Chicago
www.damngravity.com

ISBNs:
Physical: 978-1-962339-13-1
Ebook: 978-1-962339-14-8
Audiobook: 978-1-962339-15-5

Interior Layout by Bookery

Printed in The United States of America

To my son, Zane.
You are truly God's gift to Mama's and Daddy's lives.

It's such an honor and blessing to
"onboard" you to the world.

May you do greater and better things than us.

CONTENTS

THE FUNDAMENTALS

STEP 1: ESTABLISH A CROSS-FUNCTIONAL ONBOARDING TEAM

STEP 2: UNDERSTAND USER SUCCESS

STEP 3: REVERSE JOURNEY MAP TO SUCCESS

STEP 4: KEEP NEW USERS ENGAGED

STEP 5: APPLY, ANALYZE, AND REPEAT

NEXT STEPS AND CONCLUSION

APPENDIX

INTRODUCTION

"Everyone's going back to using their own tools."

I was on a call with a founder of a fast-growing B2B company who reached out to me for help. His team had spent six months trying to implement a new centralized project management tool. Despite executive buy-in, extensive training sessions, and a dedicated implementation team, teams used their own solutions: Marketing used Trello, engineering stuck with Jira, and everyone else tracked projects in various spreadsheets and docs. What was meant to unify their company had instead created chaos.

Their onboarding was clearly failing, but why? They had followed the traditional playbook: Comprehensive feature training, detailed workflow documentation, and a dedicated implementation team. They even created custom video tutorials for each department. Yet the more resources they poured into technical training, the more resistance they faced.

The root problem became clear: They had treated onboarding as a technical challenge rather than an organizational transformation. They focused on teaching features while missing the human element of change management—how teams actually work together, what motivates people to change their habits, and why cross-functional collaboration matters.

The complexity of B2B onboarding creates unique challenges. Unlike consumer apps that can delight individual users in minutes, B2B products must navigate complex organizational dynamics, technical requirements, and change management hurdles. Success demands more than just a smooth user interface or helpful tooltips—it requires a well-orchestrated onboarding experience that:

- Builds cross-functional alignment between Sales, Marketing, and IT from day one

- Maps and validates each team's path to value before the full rollout

- Creates momentum through early wins and visible progress

- Maintains engagement through a mix of automated guidance and human support

- Measures and adapts their approach based on user feedback and data

That's why complex problems like B2B onboarding require comprehensive solutions. You can't rely solely on in-product guides or tutorial videos. Successful B2B onboarding requires multiple teams working in concert: Product crafts intuitive experiences, Marketing creates educational content, Sales guides evaluation, and Customer Success ensures implementation success. You need a multi-channel strategy that combines automated guidance with human support at crucial moments.

I've seen these challenges firsthand working with hundreds of B2B companies. At ProductLed with Wes Bush (with my co-author for *Product-Led Growth*), we helped companies like Outsystems and Vidyard optimize their onboarding for product-led growth. At Appcues, a product adoption software company, we worked with teams at Bynder and Fullstory to implement effective in-product guidance. Now, through my boutique consulting firm, Delight Path, I continue helping high-growth B2B companies turn new users into successful customers.

Through my work with hundreds of B2B companies at Product-Led, Appcues, and now Delight Path, I've developed the EUREKA

Framework for successful B2B onboarding. Since I first introduced this framework in my previous book, *Product-Led Onboarding*, I've refined it specifically for high-growth B2B companies. The EUREKA Framework is an acronym for the five-step blueprint to help you build a well-orchestrated B2B onboarding experience:

1. **Establish your onboarding team:** You need to take a cross-functional approach to deliver an effective, immersive, and seamless B2B onboarding experience. In Chapters 4 through 6, I'll help you form an onboarding team.

2. **Understand user and customer success:** The best way to onboard new customers successfully is to figure out why they signed up and purchased your product in the first place. Mainly, you need to know what value means to users. In Chapters 7 through 9, I'll help you figure out how to do the User Success Canvas and Four Forces of B2B Product Adoption.

3. **REverse journey map to success.** The next step is to map out the critical path to value realization. By working backward from your customer's desired outcomes, you can identify key milestones and touchpoints needed for success. This reverse engineering approach helps eliminate unnecessary steps and focuses on what truly matters for adoption. In Chapters 10 through 12, we'll explore how to create effective journey maps that align with your customer's goals and accelerate their path to value.

4. **Keep new customers engaged.** Once you have the journey map with key customer milestones, we'll map out the hierarchy of B2B user friction and the three pillars of B2B onboarding to it. In Chapters 13 through 18, we'll

explore strategies to maintain momentum through product guidance, educational content, and human touchpoints. This includes designing engaging product tours, creating targeted help documentation, and implementing effective customer success check-ins.

5. **Apply, Analyze, and Repeat.** In the last step of the EUREKA Framework, we bring it all together. In Chapters 19 and 20, we'll explore how to implement your onboarding strategy, measure its effectiveness, and continuously improve based on user feedback and data.

By the end of this book, you'll have a proven framework for consistently turning new users into product champions who drive organization-wide adoption, reduce time-to-value, and generate expansion revenue through team-wide deployment.

WHO SHOULD READ THIS BOOK

This book is for B2B teams who:

- Have achieved product-market fit and aim to scale
- Navigate complex buying cycles and implementation processes
- Want to optimize existing onboarding programs

If you're building from scratch, start with my first book, *Product-Led Onboarding*. EUREKA focuses on scaling and optimizing established programs to drive team-wide adoption.

We'll move beyond basic signup flows or feature tutorials. We'll focus on creating comprehensive onboarding systems that transform organizations, drive team-wide adoption, and turn customers into long-term champions of your product.

WHAT THIS BOOK IS (AND ISN'T)

This book is a flexible framework that you can adapt to your organization's unique context. It's not a rigid, prescriptive solution—because every B2B product has its own complexities, user dynamics, and organizational challenges.

What you'll find here is a battle-tested approach that has helped hundreds of B2B companies improve their onboarding. The EUREKA Framework provides the structure and playbook, but you'll make it your own based on your product, market, and customers' needs.

Think of this book as your onboarding transformation guide, not a universal blueprint. The exercises, templates, and strategies are meant to be tailored to your specific situation, helping you build an onboarding experience that works for your unique context.

HOW I'VE ORGANIZED THE BOOK

The book follows a practical, implementation-focused structure:

The Fundamentals (Chapters 1–3)
We'll explore the bridge to customer success, the hierarchy of B2B user friction, and the three pillars of successful B2B onboarding that form the foundation of effective implementation.

The EUREKA Framework (Chapters 4–18)

We'll dive deep into each step of the framework:

- **E**stablish your cross-functional onboarding team (Chapters 4–6)
- **U**nderstand user success (Chapters 7–9)
- **RE**verse journey map to user success (Chapters 10–12)
- **K**eep new users engaged (Chapters 13–15)
- **A**pply, analyze, and repeat (Chapters 16–18)

Next Steps and Conclusion (Chapters 19–20)

We'll explore the concept of "everboarding" and what happens after activation.

HOW TO GET THE MOST OUT OF THIS BOOK

Key chapters end with a "Your Turn" section containing practical exercises to apply the concepts immediately in your organization. You'll find templates, checklists, and additional resources at *eurekabonus.com* to help implement these ideas quickly.

Are you ready?

All aboard!

Ramli John

Ramli "RJ" John

PS—I'd love to hear how you implement the EUREKA Framework at your company! Email me at ramli@delightpath.com or connect with me on LinkedIn (linkedin.com/in/ramlijohn).

Plus, remember to grab your bonus resources at *eurekabonus.com*. There, you'll find templates, worksheets, and examples to help you implement everything we cover in the book.

THE
FUNDAMENTALS

1

The Bridge to Long-Term Customer Success

~~~

**Nothing so undermines change as the failure
to think through the losses people face.**

William Bridges, Author of
Managing Transitions: Making the Most of Change

~~~

In *Indiana Jones and The Last Crusade* (one of my favorite movies),
Indiana Jones must cross a deadly chasm by taking a "leap of faith"
to save his father. An invisible bridge only becomes visible when he
takes that first brave step into seemingly empty space.

"You must believe, boy. You must believe," his dad tells him.

Indiana takes a breath and leaps, landing on what appears to be thin
air. The bridge materializes beneath his feet, a solid pathway hidden
by clever camouflage.

For many B2B products, that's what their customer onboarding
looks like.

Their customers made the "leap of faith" and purchased the product,
but now they're standing on an invisible bridge, unsure of their next
steps. Without proper guidance, they're left wondering if they made
the right decision.

For some products, the customer leaps, but the bridge never materializes.
They fall into the chasm of confusion, frustration, and eventual churn.

This is why product onboarding isn't just another checkbox in your customers' journey—it's the bridge between purchase and success. Like Indiana's leap of faith, customers need to trust that their decision was right. But unlike the movie, we can't rely on ancient Templar architecture. We must build that bridge ourselves.

THE BRIDGE FROM STRUGGLE TO SUCCESS

The bridge is an analogy I love to use to describe what onboarding is and why it's so important. So you'll see this again in the next chapter and the rest of the book.

When customers first encounter your product, they're on one side of a chasm. On this side is their current situation—their struggles, challenges, and pain points. On the other side lies their desired outcome—the success and value they hope to achieve with your product. Onboarding is the bridge that connects the customer's struggling moment to their desired outcome. Your goal is to guide them safely while building confidence in their decision to choose your solution.

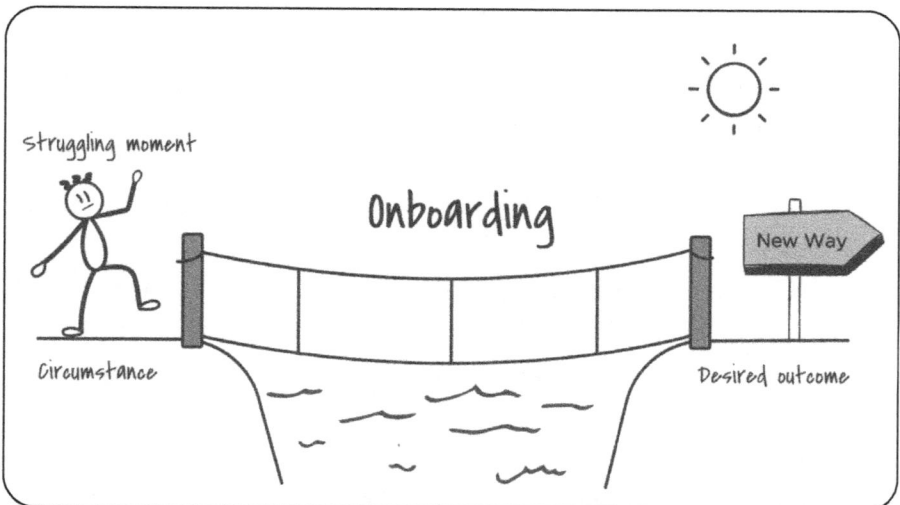

Without a sturdy bridge, customers risk falling into the chasm of fear, frustration, and confusion. They might struggle to understand how to use your product effectively, fail to see its value, or become overwhelmed by complexity. They may let their fear of change prevent them from fully adopting your product. Even worse, they might retreat to their familiar but ineffective ways of working. The result is poor adoption rates, increased support costs, and customer churn.

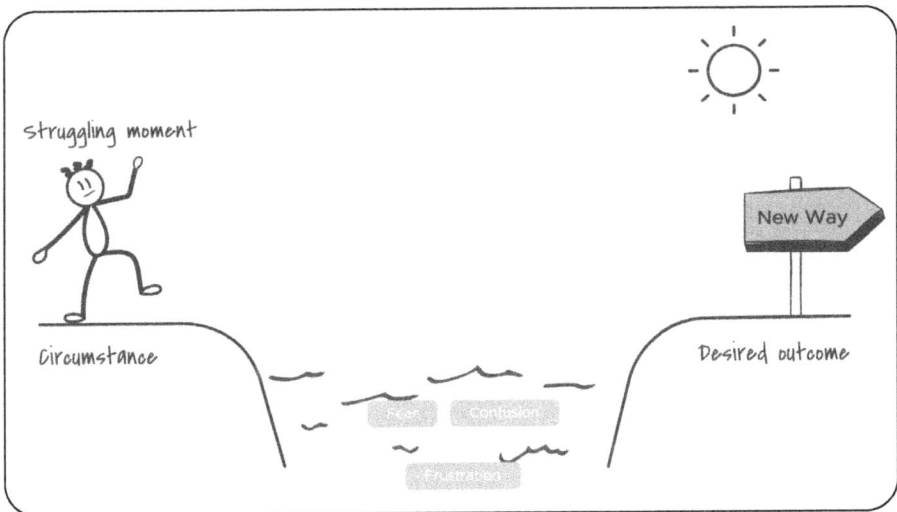

Think of onboarding as your product's first impression in action. **You're not just onboarding new customers to your product—you're guiding them through a transformation. You're helping them embrace new ways of working, thinking, and achieving their goals.**

THE IMPACT OF BAD PRODUCT ONBOARDING

The problem is that change is hard and scary. Even when customers know they need to change, the fear of disrupting established workflows or making mistakes can paralyze them. This resistance to change is one of the biggest barriers to success, leading to avoidable churn, weak internal advocacy, and reduced customer lifetime value—and it's largely preventable with proper onboarding.

1. Avoidable Churn

Poor onboarding leads to customers abandoning your product within the first few months. A 2019 study by Totango reveals that companies lose 75 percent of their new users within the first week without proper onboarding. This early churn is often due to customers feeling overwhelmed, confused, or unable to achieve their desired outcomes quickly enough.

When customers can't see the value in your product during those crucial first weeks, they're more likely to cancel their subscription and seek alternatives. Salesforce's State of Customer Report backs this sentiment up, which found that 67 percent of customers would switch products due to a poor onboarding experience.

Retention starts during the onboarding experience—and the data proves it.

ProfitWell studied about five hundred different B2B software companies. They found that customers with a positive onboarding experience were more likely to stick around than those who weren't happy with it.

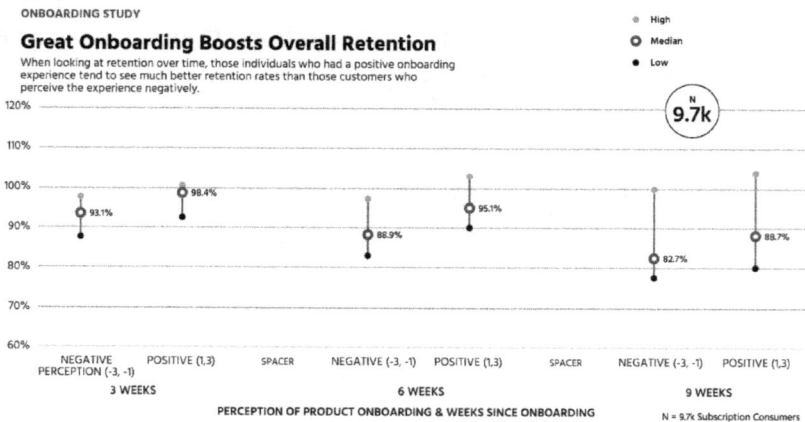

ONBOARDING STUDY

Great Onboarding Boosts Overall Retention

When looking at retention over time, those individuals who had a positive onboarding experience tend to see much better retention rates than those customers who perceive the experience negatively.

High / Median / Low

N 9.7k

		98.4%				95.1%		
93.1%				88.9%				88.7%
							82.7%	

NEGATIVE PERCEPTION (-3, -1) / POSITIVE (1,3) / SPACER / NEGATIVE (-3, -1) / POSITIVE (1,3) / SPACER / NEGATIVE (-3, -1) / POSITIVE (1,3)

3 WEEKS   6 WEEKS   9 WEEKS

PERCEPTION OF PRODUCT ONBOARDING & WEEKS SINCE ONBOARDING

N = 9.7x Subscription Consumers

InnerTrends saw similar data points: users who completed the initial onboarding process were 38 percent more likely to return one week later.

But it goes even further. Once users hit Week 12, the effects of user onboarding are even more pronounced. For those who completed Inner-Trend's onboarding process, the retention rate is almost three times higher.

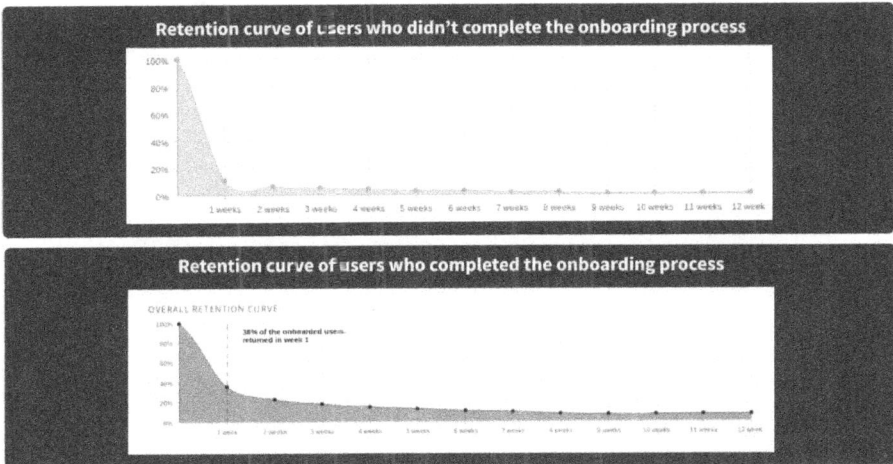

When someone first signs up for a product, they'll either love it or leave it. Those who are successfully onboarded see the value and are more likely to stick around, even years later.

2. Internal Advocacy

Poor onboarding has a devastating impact on revenue and growth for B2B companies. According to Gartner, 80 percent of B2B customers report that their implementation experience impacts their likelihood of recommending a product or service internally within their organization.

For product-led companies, this internal advocacy is crucial for expansion and growth. When customers struggle during onboarding, they're less likely to become product champions within their organization, limiting your ability to expand accounts and secure renewals.

A bottom-up growth strategy is how Slack quickly scaled. It starts with one person (let's say Mary) using it with a colleague. Mary loves Slack's productivity benefits so much that she started using it for her team. Pretty soon, the whole company is on Slack.

In this product-led growth approach, end users find solutions independently and advocate for them to their bosses.

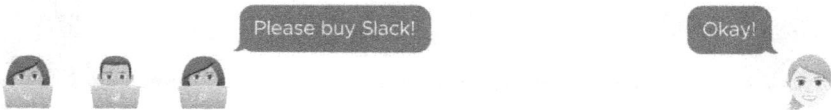

This bottom-up adoption model relies heavily on successful onboarding experiences. When users like Mary encounter friction or confusion during their initial product experience, they're less likely to champion the solution to others. Effective onboarding creates confident, enthusiastic users who naturally become product advocates.

For sales-led companies, this internal advocacy is necessary to secure long-term contract renewals and navigate complex buying committees. When customers struggle to adopt your product effectively, they're less likely to see its full value, making it harder to justify renewals.

I've been in situations where the company's leadership team purchased a product they hoped their employees would use to increase productivity, reduce cost, or drive new business. However, employees resist adopting these new tools without proper onboarding and training, viewing them as unnecessary disruptions to their existing workflows. This resistance creates a gap between leadership's vision and implementation, resulting in unrealized potential and lost revenue. The key to bridging this gap lies in an onboarding experience that demonstrates immediate value to end users while addressing their pain points and concerns.

3. Customer Lifetime Value

Poor onboarding directly impacts the long-term value customers bring to your business. New customers who struggle to adopt your product effectively are less likely to upgrade to premium features, expand their usage, or maintain long-term loyalty. According to Salesforce's State of the Connected Customer report, companies that invest in comprehensive onboarding programs see an average increase of 34 percent in customer lifetime value, primarily because well-onboarded customers stay longer and spend more over time.

Having a healthy LTV is critical because most B2B companies often require customers to stick around for twelve to eighteen months before they can recoup the customer acquisition costs (CAC). This metric is called the CAC Payback Period, which measures how long it takes for a company to recover its investment in acquiring a customer. When customers churn before this period ends due to poor onboarding, companies not only lose future revenue but also fail to recoup their initial investment.

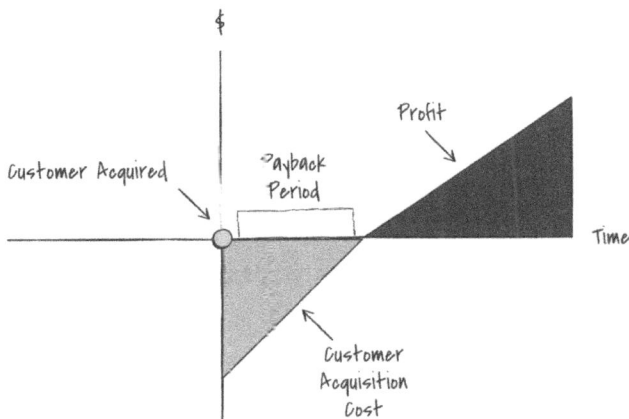

It's not just about paying back the investments you've made in customer acquisition. Effective onboarding also creates opportunities for upselling and cross-selling, which significantly boost LTV. According to a study by Totango, companies that invest in customer onboarding see an average increase of 32 percent in annual recurring revenue from existing customers through expansions and upgrades.

When customers fully understand and adopt your product's core features, they're more likely to explore and upgrade to premium tiers, add more users, or purchase additional features. This expanded usage translates into higher revenue per customer and stronger relationships.

This opens up the opportunity to achieve a **net negative churn rate,** where revenue growth from existing customers exceeds losses from churn. Companies that achieve net negative churn often see their customer base naturally expand without requiring additional sales efforts, leading to increased profitability as revenue grows without proportional increases in acquisition costs. For example, if a company loses 10 percent of its revenue to churn but gains 15 percent through expansions, it achieves a net negative churn rate of -5 percent.

And it all starts with the onboarding.

THE CUSTOMER'S PERSPECTIVE

Bad onboarding undermines your business and creates significant challenges for customers, including buyer's remorse, missed opportunities, and wasted resources.

Buyer's Remorse

Poor onboarding often leads to **buyer's remorse**, where customers regret their purchase. This is especially common with luxury items like designer handbags or expensive watches. In B2B contexts, buyer's remorse can be even more severe, as decision-makers worry about wasting company resources or damaging their professional reputation and career. According to a 2024 Gartner study, 60 percent of B2B buyers express some form of buyer's remorse after major software purchases. **Effective onboarding reduces buyer's remorse by validating the customer's decision early and often.** This means demonstrating quick wins, highlighting value milestones, and reinforcing the benefits they sought when choosing your product. When customers can see tangible progress toward their goals within the first few days or weeks, they're more likely to feel confident in their purchase decision and commit to long-term adoption.

Missed Opportunity to Ride the Excitement

Customers are most excited and motivated to make a change at the moment of purchase. They've just committed to your solution and are eager to see results. Poor onboarding wastes this valuable momentum and emotional high.

When customers don't have clear next steps right after the purchases, they lose motivation and enthusiasm for implementation. This initial excitement is a critical window for establishing strong customer engagement and relationships. Without proper guidance during this phase, customers may become frustrated and disengage, making it harder to rekindle their interest later. The momentum from their purchase decision needs to be channeled into productive actions that demonstrate immediate value and build confidence in their choice.

It's why I'm a big proponent of getting customers to a "quick win" that they can achieve within the same day of purchase or signup. I call these **Same-Day Wins** (which we'll cover more in detail in Chapter

11: Mapping The Onboarding Success Roadmap). Other examples of Same-Day Wins from other companies are:

- Appcues: Build and design a product tour
- Airbnb: Save a listing to a "wishlist"
- Freshdesk: Reply to and close a "dummy" support ticket
- HubSpot: Add your first contact
- Enterprise: Schedule your kickoff call

Wasted Time and Resources

Every delay in the onboarding process represents lost productivity and customers' resources. When teams spend excessive time troubleshooting setup issues or trying to understand basic functionality, they're not focusing on their core business objectives. This is particularly costly for enterprise customers, where multiple stakeholders and departments may be involved in the implementation.

According to a study by Forrester Research and Airtable, employees spend an average of 2.4 hours per day searching for information and navigating new software systems during implementation periods. Poor onboarding results in nearly 30 percent of their workweek spent on non-productive activities.

By building a streamlined onboarding process, you're honoring the trust customers have placed in you and your product. They've invested their time in vetting your solution, and now it's your responsibility to guide them with confidence and care. You're removing the burden from customers to figure everything out on their own, assuring them they're in good hands throughout their journey. This human-centered approach shows customers that you understand their challenges and are committed to their success from day one.

THE TRUST MULTIPLIER

There's a saying that it's not about business-to-business or business-to-consumer; it's human-to-human. While this may sound cliché, it highlights an essential truth about customer relationships.

Trust is the foundation of all successful business interactions, and onboarding is your first real opportunity to build that trust. As Stephen Covey writes in *The Speed of Trust*, "Trust is equal parts character and competence. You can look at any business failure, and it's always a failure of one or the other." This perfectly describes the onboarding challenge: you must demonstrate both your product's competence and your organization's character in supporting customer success.

When customers feel supported and valued during onboarding, they're more likely to become advocates for your brand. This creates a multiplier effect, where satisfied customers not only stay longer but actively promote your solution to others.

Building trust starts with understanding and anticipating your customers' challenges. The more prepared you are to address common friction points, the more confidence customers will have in your ability to guide them. In the next chapter, we'll examine the three biggest barriers to successful B2B onboarding and how to overcome them.

2

The Hierarchy of B2B User Friction

~~~

My goal is to simplify complexity. I just want to build stuff that really simplifies our basic human interaction.

Jack Dorsey, Co-founder of Twitter and Square

~~~

At my best friend's bachelor party, I went bungee jumping for the first and last time. He insisted on doing it before getting married—I guess he wanted to practice taking the plunge. Standing at the edge of that platform, my heart racing, I realized there were multiple layers to my hesitation. Yes, there was the physical challenge of actually jumping. But before I even got there, I had to overcome my lifelong fear of heights. And perhaps hardest of all, I had to deal with the social pressure—my friends were watching, and I didn't want to back out after talking big about doing it. I took the leap, and it was exhilarating.

As a product growth consultant who has worked with hundreds of B2B companies, I've noticed that product onboarding faces similar multi-layered challenges. Complex approval processes, security reviews, and team coordination create more significant obstacles than any in-product issues.

Through my work, I've identified four key challenges that set B2B onboarding apart. After examining these foundational challenges, we'll explore how they manifest in three distinct levels of friction that every onboarding team must overcome.

WHAT MAKES B2B ONBOARDING CHALLENGING?

B2B onboarding has at least four unique challenges that set it apart from B2C products:

1. B2B products are more complex.

Successful B2B onboarding often requires integrating the product with existing systems and workflows. Users need to understand not just the product features but how these features fit into their broader business processes. For example, when implementing Salesforce, companies must consider how it integrates with their email systems, marketing automation tools, customer support platforms, and existing databases. Users need training on data entry protocols, pipeline management, and customized reporting. The implementation process often involves multiple stakeholders across departments, each with its own requirements and workflows.

2. B2B products involve multiple users, decision-makers, and stakeholders.

B2B products typically serve multiple users, decision-makers, and stakeholders within an organization—each with different needs and priorities. For example, in a CRM implementation, sales managers focus on reporting features, sales representatives learn data entry protocols, and IT teams evaluate security settings. Coordinating multiple teams to transition from their existing tools and processes often creates delays and bottlenecks.

Like building a bridge, B2B onboarding requires moving entire teams from their current tools and processes to new ones—not just helping a single user crossover.

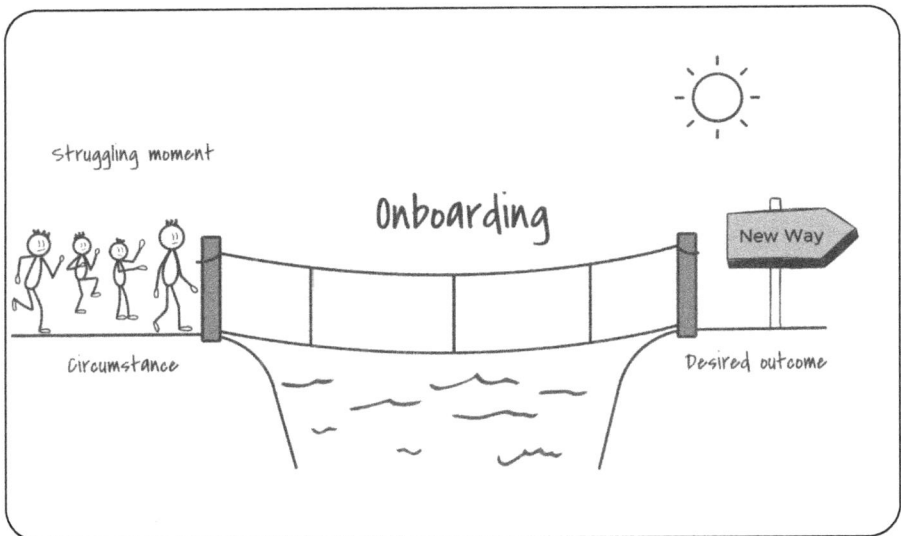

3. B2B products have longer time-to-full-value.

Unlike B2C apps that deliver immediate value, B2B products often need weeks or months before organizations see their full value. At Appcues, customers typically take fourteen to thirty days just to reach their first activation moment—publishing their first user flow. Technical requirements, team coordination, and approval processes all extend this timeline.

That's why getting users to a quick win or Same-Day Win is crucial. These early victories help maintain momentum during the longer implementation journey. (We'll explore this further in Chapter 11: Mapping the Onboarding Success Roadmap.)

4. B2B products have a wide range of use cases.

B2B products often evolve to serve diverse business needs across different industries and company sizes. Take Slack—it started as an internal tool for a gaming company but grew to serve teams in tech, media, healthcare, and education. Each industry brings unique requirements, from compliance needs to workflow patterns.

Various use cases make one-size-fits-all onboarding impossible. A healthcare company needs different security protocols and workflow configurations than a marketing agency using the same tool. The onboarding process must flex to accommodate these varying needs.

THE HIERARCHY OF B2B USER FRICTION

These four challenges reveal a crucial insight: **The biggest barriers to B2B adoption often lie *outside* the product itself.** Yet in my work with hundreds of B2B companies, I've noticed most teams focus solely on basic product friction, like:

- Setting up their account
- Uploading their data into the product
- Installing your code snippet into their production

The more challenging barriers to successful B2B onboarding often involve organizational change and team dynamics:

- Getting stakeholder buy-in for process changes
- Securing budget approval from management
- Coordinating team-wide training and adoption
- Integrating with existing workflows
- Managing organizational resistance to change

To better understand and address these challenges, I categorize B2B user friction into three distinct levels:

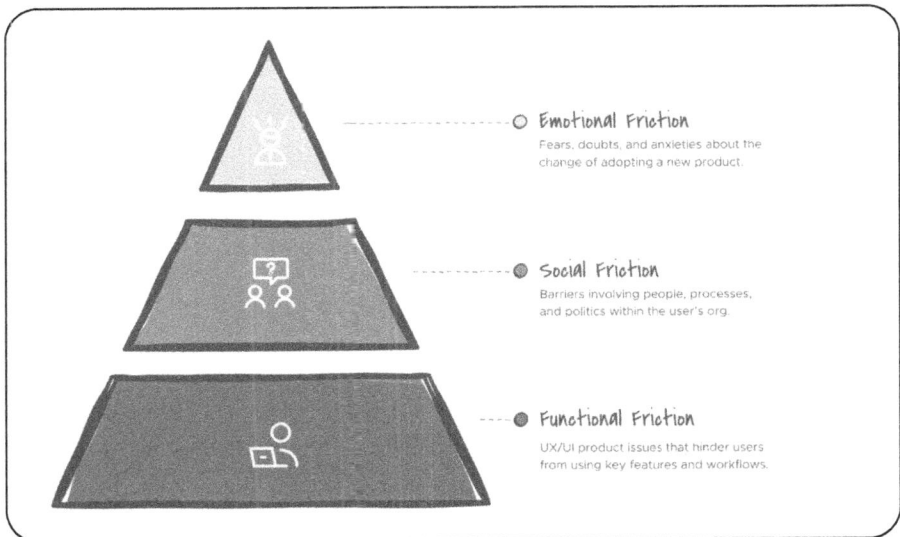

Emotional Friction
Fears, doubts, and anxieties about the change of adopting a new product.

Social Friction
Barriers involving people, processes, and politics within the user's org.

Functional Friction
UX/UI product issues that hinder users from using key features and workflows.

Emotional friction consists of the psychological barriers that prevent users from fully embracing new software solutions. These barriers include fears about making the wrong purchasing decision, anxiety about learning new systems, and concerns about how changes might affect their job security or professional reputation.

Social friction is the organizational and interpersonal barriers that arise during product adoption. It encompasses everything from internal politics and competing departmental priorities to the challenges of coordinating multiple stakeholders and managing cross-team dependencies. Social friction often appears as resistance to new workflows, difficulty securing buy-in, and challenges aligning different teams' objectives.

Functional friction is the fundamental usability and technical barriers users encounter when trying to use your product. This includes interface issues like confusing navigation and unclear error messages, as well as technical challenges like slow performance, integration difficulties, and data migration problems. While these are often the most visible barriers, they're typically the easiest to identify and address.

Let's take a deeper look at each, starting from the lowest level of B2B friction:

1. Functional Friction

Most teams start by addressing functional friction—the basic usability barriers in their product. While important, these issues are often the easiest to identify and fix using analytics tools like Fullstory, Amplitude, Posthog, or Mixpanel. Most content and resources around reducing onboarding friction focus heavily on this level (even my bestselling book, *Product-Led Onboarding*).

Common functional friction points for B2B products include:

- **Technical Setup:** Barriers related to technical configuration, including API setup and system compatibility issues

- **Navigation**: Confusing UI elements, unclear menu structures, and poor information architecture that make it difficult for users to find features or understand their progress

- **Knowledge Gaps**: Confusion about features and best practices that prevent effective product use

- **Data Migration**: Challenges moving existing data while maintaining integrity and relationships

- **Integration**: Difficulties connecting with existing tools and workflows in the tech stack

Take Salesforce's complex interface: New users often struggle with basic tasks like creating contacts or logging activities. The platform's extensive customization options mean simple actions can become overwhelming—there might be dozens of required fields using company-specific terminology, multiple record types to choose from, and nested menu structures that hide common functions.

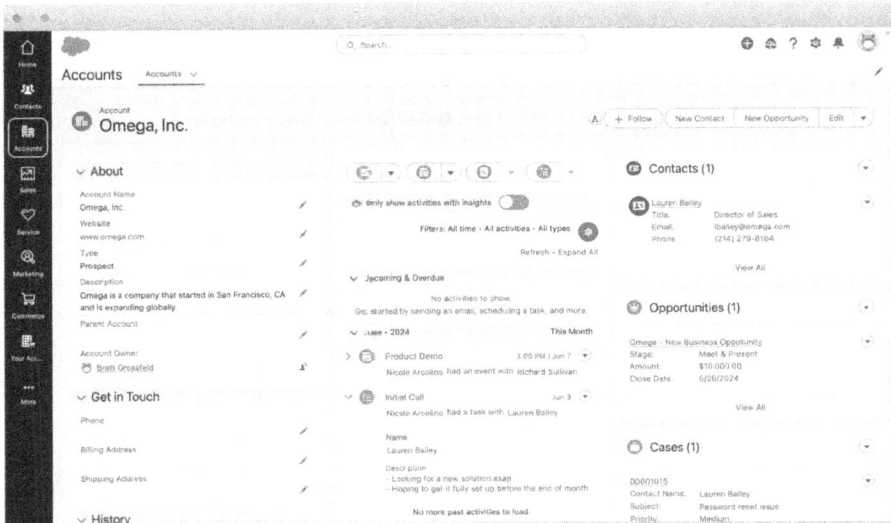

Even seasoned sales professionals can get lost trying to find where to log a simple phone call or update a contact's information. But solving these UI challenges alone won't guarantee successful adoption—you must also address the social and emotional barriers users face.

Focusing solely on functional friction misses the bigger picture. Even with a perfectly designed interface and smooth technical implementation, B2B products can still fail to achieve widespread adoption if they don't address the more complex social and emotional barriers that users face.

2. Social friction

Social friction encompasses the interpersonal and organizational challenges that arise during B2B product adoption. It can manifest internally with onboarding teams or externally within the customer's organization.

Within your team, it appears as misaligned goals and communication gaps between teams managing different parts of onboarding:

- The Marketing team manages onboarding emails.
- The Sales team sets initial customer expectations.
- The Product team controls in-product experiences.
- The Customer Success team handles one-on-one interactions.

Without coordination, these teams can create a disjointed experience that frustrates users. For example, sales might promise features that aren't ready, while marketing sends emails about capabilities that don't match the product's current state.

In customer organizations, social friction surfaces when stakeholders have competing priorities or teams resist new workflows. For product-led companies, it happens when individual users struggle to gain organizational support and budget approval. They may love your product but face resistance when expanding usage across their team.

For sales-led companies, social friction in the onboarding occurs when departmental silos and competing priorities slow down or derail successful adoption. Even with executive buy-in, individual teams may resist implementation if they don't see clear value or feel threatened by change.

Common social friction patterns in B2B products include:

- **Coordination Friction**: Challenges aligning multiple stakeholders and managing dependencies
- **Change Management**: Organizational resistance to adopting new solutions
- **Resource Constraints**: Limited time, attention, and personnel for implementation
- **Stakeholder Alignment**: Conflicting priorities between different organizational levels
- **Process Changes**: Difficulties standardizing workflows across teams

For Salesforce, social friction often manifests when sales teams resist adopting new processes, even after management mandates the platform's use. Individual salespeople may continue using their spreadsheets or legacy systems, creating data silos and reducing the platform's effectiveness.

3. Emotional Friction

Emotional friction is the psychological barrier users face when adopting new products. It's the fear of making the wrong buying decision, anxiety about learning new systems, and concerns about job security.

There's a reason why "nobody ever got fired for buying IBM"—choosing established vendors feels emotionally safer than risking career reputation on newer solutions. These anxieties often hide behind rational-sounding objections about features or pricing.

Common emotional friction patterns include:

- **Value Perception:** Difficulty justifying investment and demonstrating ROI

- **Cultural Resistance:** Organizational-level resistance to change, often from leadership who are invested in existing systems ("This is how we've always done it")

- **Workflow Anxiety:** Individual-level fear about disrupting personal work routines and learning new processes

- **Implementation Doubt:** Uncertainty about successfully deploying the solution

- **Trust Issues:** Concerns about data security and vendor reliability

For Salesforce, emotional friction often manifests when sales teams resist adopting new processes (change management) even after management mandates the platform's use.

Individual salespeople may continue using their spreadsheets or legacy systems, creating data silos and reducing the platform's effectiveness.

HOW DO YOU OVERCOME EACH LEVEL OF FRICTION?

Mapping the three types of friction to Maslow's Hierarchy of Needs, functional friction corresponds to basic physiological and safety needs. Social friction aligns with belongingness needs, while emotional friction maps to psychological needs like security and esteem.

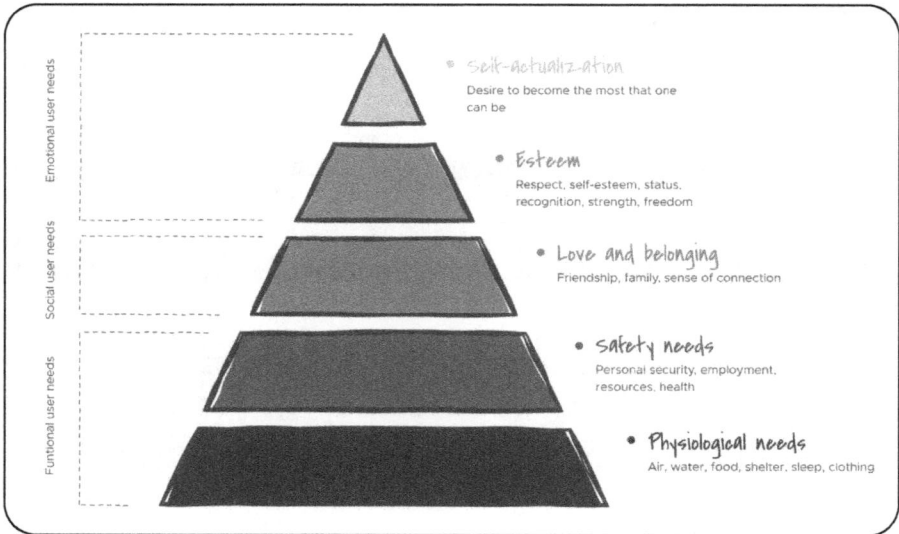

Each level requires different solutions:

- Functional friction: Better UI/UX design and documentation
- Social friction: Educational content and peer support networks
- Emotional friction: High-touch human intervention and executive sponsorship

In Chapters 13 to 16, we'll explore comprehensive solutions for addressing each level of friction. We'll start by mapping and auditing friction points (Chapter 13) and exploring the behavioral psychology principles that drive successful onboarding (Chapter 14). Then, we'll examine all three pillars of successful onboarding—in-product guides, educational content, and human interaction—in detail (Chapter 15). Finally, we'll put these insights into action through the Friction-to-Action Workshop (Chapter 16).

But first, in the next chapter, we will examine the three pillars of successful B2B onboarding.

3

The Three Pillars of Successful B2B Onboarding

~~~~~~

An aha moment is not a happy ending—it's an open doorway, one you have to choose to walk through.

Alicia Keys

~~~~~~

A **few years ago,** I picked up a form of rock climbing called bouldering as a hobby. Unlike traditional climbing with ropes, bouldering involves scaling shorter walls using your hands, feet, and a crash pad below. I got a crash course from a friend who works at the gym. I learned to fall without hurting myself and stabilize my core while climbing. The most important lesson was about the three contact points rule—always try to maintain three points of contact with the wall, whether two hands and a foot or two feet and a hand. It helps maintain balance and prevents dangerous falls.

Similarly, high-growth B2B companies can't rely on a pure product-led approach to onboarding. This may work for early-stage startups, where focusing on one use case for a market segment makes sense. But, as companies grow past product-market fit, they start serving diverse customer segments with varying needs. As discussed in the previous chapter, social and emotional friction in onboarding becomes more challenging than product friction.

Sales-led companies face a similar challenge. As they scale, they can't rely on human-led onboarding, which becomes too resource-intensive and costly. They need to implement automated and one-to-many onboarding tactics to supplement their high-touch onboarding.

THE THREE PILLARS OF B2B ONBOARDING

That's why, to onboard customers effectively, high-growth B2B organizations need three key onboarding pillars:

1. In-product guides

These are interactive walk-throughs, tooltips, and contextual help that direct users through key workflows and features. This automated guidance helps users navigate the product independently while reducing support tickets.

2. Educational content

These are courses, video tutorials, webinars, and knowledge-base articles that provide comprehensive product training and best practices to quickly get users up to speed. These resources enable self-paced learning and serve as ongoing reference materials for teams.

3. Human touchpoints

These are human touchpoints through customer success managers, onboarding specialists, and technical support teams who provide strategic guidance, troubleshooting assistance, personalized recommendations, and support for company-wide product adoption.

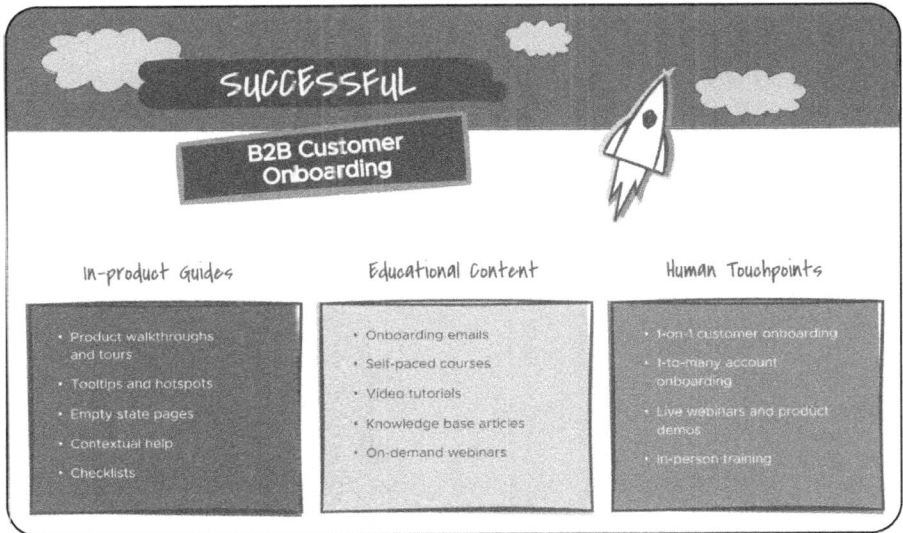

These three pillars work together to create a comprehensive onboarding experience. In-product guides provide immediate assistance, educational content builds deeper product knowledge, and human touchpoints ensure strategic alignment and success for enterprise customers. In Chapter 15, we'll explore these pillars in detail, including specific tactics, implementation strategies, and examples of how leading companies use them effectively.

Best-in-class product-led B2B companies like Slack, Asana, and HubSpot exemplify this hybrid approach. They combine intuitive in-product experiences with comprehensive learning centers and dedicated customer success teams.

For instance, Slack's interactive tutorials and templates help new users quickly understand core functionality and experience the product's value.

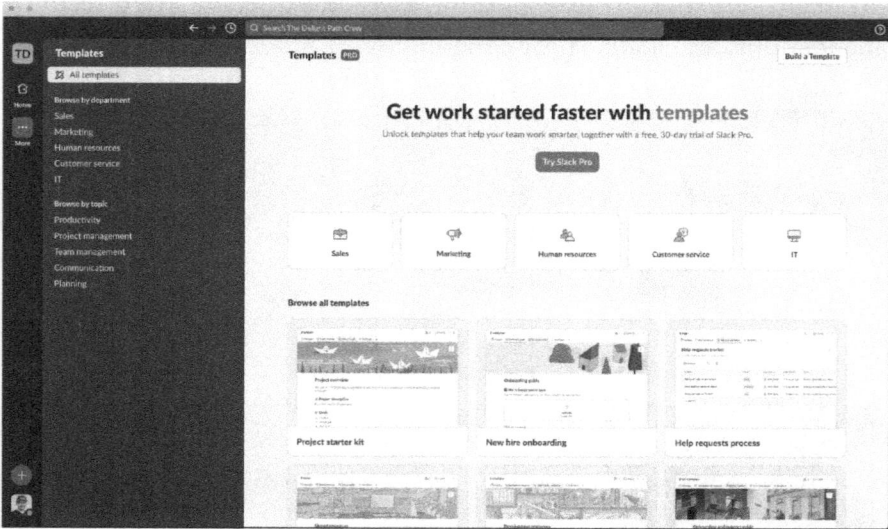

Their Slack Certified program provides deeper product mastery for power users and Slack administrators of enterprise customers who need comprehensive knowledge of advanced features and integrations.

Slack Certified

Grow your Slack skills and prove your knowledge to the world.

Slack Skills

Take courses to earn badges and become a specialist in Slack.

Earn badges →

Slack Certified Admin

For Slack admins at organizations of all sizes, unlock the full functionality of Slack and verify your proficiency as a Slack Certified Admin.

→

Slack Certified Consultant

Prove you have the skills to implement Slack and set clients up for long-term success.

→

In their Slack Success Hub, they have detailed documentation, best practices, and implementation guides for different team sizes and use cases.

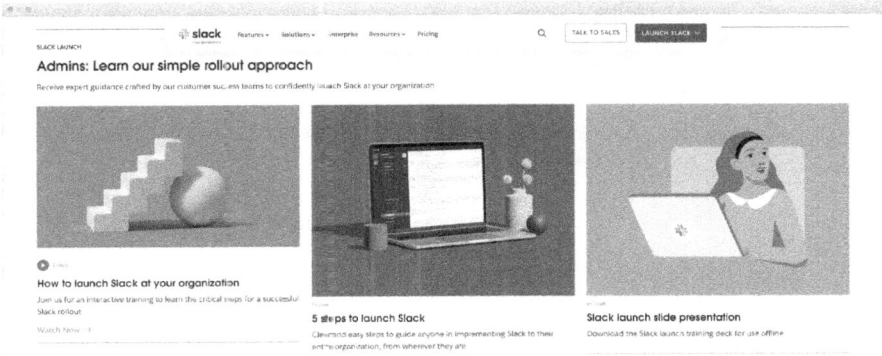

Meanwhile, customer success teams can focus on strategic initiatives, ensure enterprise-wide adoption, and maximize customer value through personalized consultation and support.

THE FOUNDATION OF SUCCESSFUL B2B ONBOARDING

Before you build any structure like a sturdy bridge, you need a solid foundation. In onboarding, those are product and user engagement data. For you to guide and orchestrate customers through the product onboarding experience, you'd need three critical pieces of information:

1. What actions have they already taken?

2. What are the next steps they need to take to move them along the onboarding journey?

3. What's the most effective way to guide them to those next steps?

Without knowing where users are in their journey and what guidance they need, you can't effectively orchestrate their onboarding experience. That's why you need a data foundation before effectively deploying the three pillars of B2B onboarding.

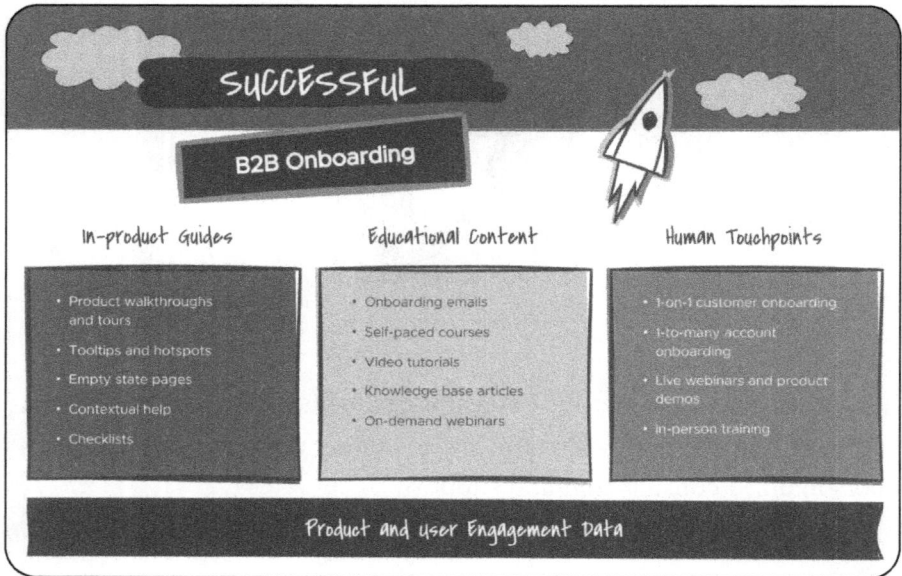

A robust data foundation enables personalized guidance and proactive support throughout the onboarding journey. This requires three essential components: real-time behavior tracking to capture user engagement, a unified customer data store that combines product usage with CRM data, and cross-tool integration that connects your analytics, support, and marketing platforms. This data infrastructure allows you to:

- Personalize the onboarding experience based on user roles and goals.
- Trigger contextual guidance at the right moments.
- Measure and optimize activation rates across customer segments.

- Identify where customers are getting stuck and proactively address their needs.

- Deploy targeted interventions such as in-product guides, educational content, or human outreach when users show signs of struggling.

So, what's the best product and user engagement data tool?

I'm cautious about suggesting an ideal onboarding tech stack, even though I frequently get asked this question. Every company's needs are unique, and the right tools depend on factors like product complexity, customer segments, and team resources.

In Step 4 of the EUREKA framework, Keep New Customers Engaged, we'll further explore the different tools that enable each of the three B2B onboarding pillars. For now, instead of focusing on specific tools, I recommend starting with your onboarding strategy (which will get into next) and then selecting tools that support those objectives.

Regardless of your tech stack, tool integration is crucial for a complete view of the customer journey. Well-integrated tools enable targeted guidance and align your GTM teams around customer needs, ensuring consistent support across all touchpoints. (We'll explore cross-functional coordination in Chapter 4.)

ORCHESTRATING THE ONBOARDING JOURNEY

So, how do you know the right mix of these three pillars for your customers?

Let me bring up another analogy with rock climbing. There are different grades of climbing routes, from beginner-friendly to highly challenging. Bouldering goes from V0 to V16, with V0 being the easiest and V16 being nearly impossible. With V0, the holds are large, easy to grip, and closer together, making it perfect for beginners to learn proper technique and build confidence. V16 routes require incredible strength, precision, and years of experience even to attempt. It requires you to break the three-point rule and make leaps to the next rock.

Like climbing routes, you need to assess each customer's "grade" and direct them to the right path. If they come from a self-serve experience, the product will be their primary guide, supplemented by educational content and human interaction. For enterprise customers with complex use cases, human touchpoints become crucial while still leveraging product guides and resources.

The best onboarding experience feels like a well-orchestrated journey where each pillar seamlessly supports the others. Product guides provide immediate assistance when users need it, educational content fills knowledge gaps and builds expertise, and human touchpoints ensure strategic alignment and onboarding coordination.

This orchestration must be carefully mapped across each stage of the onboarding journey. According to Reforge's framework, there are three critical stages of activation:

1. **Setup Step**: Customers complete essential configurations and initial setup tasks to establish a foundation for using the product.

2. **Value or "Aha" Step**: Customers experience meaningful outcomes or "aha moments" that demonstrate the product's core value.

3. **Habit Step**: Customers develop regular usage patterns and adopt the product into their workflow, leading to sustained engagement.

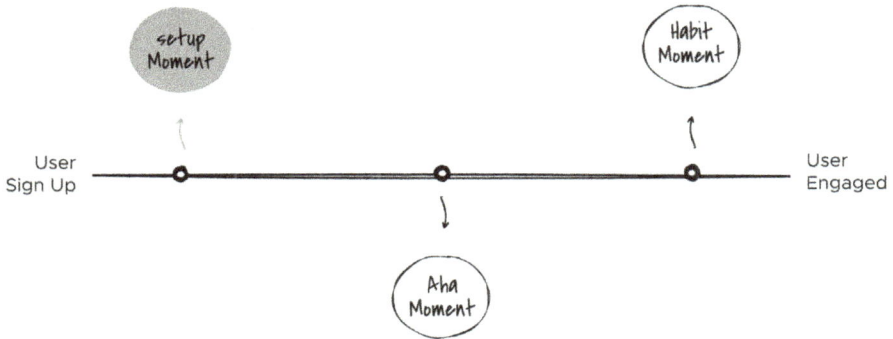

Most B2B companies struggle to orchestrate these elements effectively. While teams understand the importance of coordinating product guidance, educational content, and human touchpoints, they often struggle with the "how"—how to balance automation with human touch, scale personalized experiences, and maintain quality across all touchpoints.

THE EUREKA FRAMEWORK

That's why I developed the EUREKA Framework. This systematic approach helps teams move beyond understanding what makes good onboarding to implementing it effectively. This five-step blueprint for building well-orchestrated B2B onboarding experiences guides you through:

- **E**stablish your onboarding team
- **U**nderstand user and customer success
- **RE**verse the journey map to success
- **K**eep new customers engaged
- **A**pply, analyze, and repeat

| Establish a Team | Understand Success | Reverse Journey Map | Keep Users Engaged | Apply Analyze and Repeat |

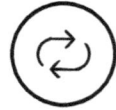

This framework will help you coordinate cross-functional teams, design personalized journeys, and deliver the right mix of product guidance, educational content, and human support at each stage of the customer journey. We'll explore each step in detail in the next five sections, starting with how to build and align your onboarding team.

EUREKA ACTION ITEMS: ASSESS YOUR ORGANIZATION

Before diving into the EUREKA Framework, it's important to understand where your organization currently stands in its onboarding journey. I've developed the B2B Onboarding Maturity Assessment (available at *eurekabonus.com*) to help you evaluate your current capabilities and identify clear paths for improvement. This assessment examines your proficiency across all three pillars—product guides, educational content, and human touchpoints—to determine which of the following three maturity stages you're at.

Most B2B companies evolve through these stages as they scale, gradually moving from reactive problem-solving to well-orchestrated experiences. Let's examine each stage in detail to help you identify where you are and what's needed to level up:

Level 1: Scattered Onboarding

At this stage, organizations excel at one primary onboarding approach—typically either high-touch or self-serve. While various onboarding components exist, they operate independently and often create a fragmented user experience due to a lack of coordination between teams.

Characteristics:

- Teams work in silos with minimal coordination
- Support-driven approach to user education
- Basic product analytics tracking
- Ad hoc documentation and guides

Three Pillars Implementation:

- In-product guides: Basic tooltips and product tours
- Educational content: Standard documentation and FAQs
- Human touchpoints: Primarily reactive support

Data Foundation:

- Limited usage tracking
- Fragmented customer data across tools
- Basic success metrics

Primary Focus:

- Addressing functional friction
- Solving immediate user problems
- Basic feature adoption

Level 2: Segmented Onboarding

At this stage, organizations develop distinct onboarding paths based on customer segments and needs. While teams create structured approaches with dedicated resources for different user types, divided ownership of these segments (e.g., self-serve versus enterprise customers) often leads to coordination challenges and inconsistent experiences across customer journeys.

Characteristics:

- Segmented onboarding paths based on customer size and complexity
- Blend of self-serve and high-touch strategies
- Established playbooks for different customer types
- Initial cross-functional coordination efforts
- Resource optimization across channels

Three Pillars Implementation:

- In-product guides: Segment-specific onboarding flows and feature education
- Educational content: Mix of automated resources and live training sessions

- Human touchpoints: Tiered support model based on customer needs and value

Data Foundation:

- Segment-based analytics and tracking
- Customer health scoring system
- ROI measurement for different approaches
- Usage patterns by customer segment

Primary Focus:

- Finding the optimal mix of touchpoints for each segment
- Balancing automation with human interaction
- Scaling operations without sacrificing quality
- Improving resource allocation efficiency

Level 3: Seamless Onboarding

At this stage, cross-functional teams work in harmony to deliver and continuously optimize the onboarding experience. Data-driven insights drive improvements across all touchpoints, creating a seamless journey that adapts to different customer needs.

Characteristics:

- Deep cross-functional alignment
- Data-driven decision-making
- Continuous optimization
- Proactive user enablement

Three Pillars Implementation:

- In-product guides: Contextual and personalized experiences

- Educational content: Multi-format learning paths tailored to user roles

- Human touchpoints: Strategic enablement and proactive guidance

Data Foundation:

- Integrated data across all tools

- Predictive analytics capabilities

- Real-time insights and automation

Primary Focus:

- Addressing all three friction types (functional, social, emotional)

- Enabling organizational transformation

- Driving team-wide adoption

Moving Up the Maturity Model

While each organization's journey is unique, the progression remains consistent: Establish foundational processes, scale with the right mix of approaches, and then optimize through integration and personalization. To help you assess your current stage and identify the next steps, complete the B2B Onboarding Maturity Assessment at *eurekabonus.com*.

STEP 1

Establish a Cross-Functional Onboarding Team

Establish a Team | Understand Success | Reverse Journey Map | Keep Users Engaged | Apply Analyze and Repeat

4

The Hidden Onboarding Friction

~~~~~~

**The strength of the team is each individual.
The strength of each member is the team.**

Phil Jackson

~~~~~~

One of my favorite games growing up was "Pass the Message" (known elsewhere as "Telephone" or "Chinese Whisper"). A staple at Filipino birthday parties, it's a simple game in which kids sit or stand in a line. The party host whispers a secret message into the first child's ear—usually something long or complicated.

The original message might be something like, "Every morning before school, I eat three yellow bananas and two sweet mangoes while watching cartoons with my little sister." That child would then whisper what they heard to the next person, and so on down the line. By the time the message reached the last child, it would be hilariously distorted. The final version might come out as "Every monkey at the zoo eats bananas and dances with mangoes while watching TV!" Mark, my younger brother, once confessed to me that he'd deliberately change the message when it was his turn, making the final result even more outrageous.

While this communication breakdown is hilarious as a game, it perfectly illustrates the hidden friction that plagues many onboarding experiences: departmental misalignment. When Marketing, Sales,

Product, and Customer Success teams aren't in sync, they create a frag-mented experience where—just like in the children's game—the core message gets distorted as it passes between departments. Each team communicates differently with customers about the product's value, features, and path to success, leaving users confused and frustrated. This miscommunication between departments is the hidden friction that silently undermines onboarding success.

The only way to build a well-orchestrated onboarding journey is with a cross-functional approach. **Onboarding is not a product problem or a customer success problem; it's a business problem that requires perfect alignment across all customer-facing teams.**

That's why establishing alignment is the first step in the EUREKA Framework. Before we dive into specific onboarding tactics or user experiences, we need to address the organizational foundation that makes great onboarding possible. This chapter focuses on how your company's internal structure and communication patterns directly shape the customer's onboarding journey. We'll examine how mis-alignment between teams creates hidden friction that undermines onboarding success and explore the key warning signs that indicate your organization's structure itself may be the biggest obstacle to smooth customer onboarding.

THE FIVE SYMPTOMS OF ONBOARDING MISALIGNMENT

Miscommunication and misalignment between departments are the hidden frictions that undermine onboarding success. While most teams focus on improving their individual parts of onboarding, the gaps between teams often cause the biggest problems.

When departments aren't aligned, it becomes exponentially harder to address the three core types of friction we discussed in Chapter 2: functional (completing tasks), social (product advocacy), and emotional (user confidence).

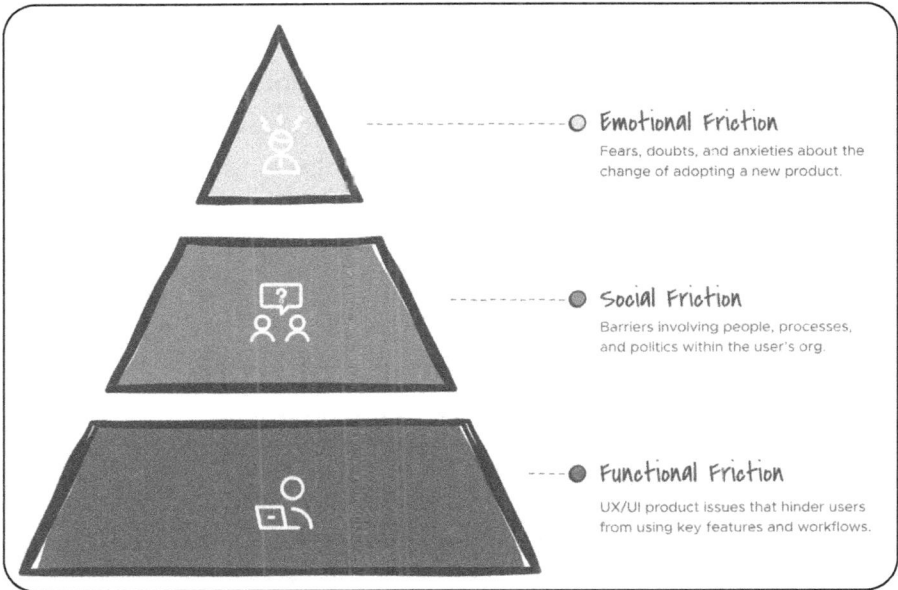

Each team tries to help in its own way to overcome these friction points—Product simplifies the interface, Customer Success provides hands-on support, and Marketing shares success stories. However, without coordination, these well-intentioned efforts often create a jarring customer experience.

After working with hundreds of B2B companies, I've identified five key symptoms of an onboarding coordination problem:

1. No clear definition of onboarding success

Without a shared definition of success, your company is like a boat in which each department is rowing in a different direction—Sales toward closing deals, Product toward feature adoption, Customer Success toward outcomes, and Marketing toward engagement metrics.

The result? Customers receive mixed signals about what success looks like, making their journey feel chaotic and confusing. Just like in our game of Pass the Message, the original meaning gets lost as it travels between teams.

Here's a quick test: Ask your teams how they define successful onboarding. If you get different answers or blank stares, you have an alignment problem. We'll explore how to fix this in Step 2: Understand User Success, but for now, recognize that this misalignment is often the root cause of fragmented onboarding experiences.

2. No single source of truth for customer data

When there's no single source of truth for customer data, you won't know if you've successfully onboarded a customer. The problem is that teams often use customer data across multiple tools—Sales uses the CRM, Product tracks usage in analytics, Support manages tickets in the help desk, Implementation keeps notes in project management tools, and Marketing tracks engagement in email platforms.

Fragmented data leads to duplicated documentation and missed opportunities to help customers during onboarding. With unified data, teams could trigger targeted emails when users get stuck or alert Sales when accounts need attention. Instead, like our game of Pass the Message, important signals get lost as information travels between systems.

3. Unclear handoffs between teams

Every customer journey involves multiple handoffs between teams—Marketing to Sales, Sales to Implementation, and Implementation to Customer Success, with Support weaving throughout. Without clear ownership of these transitions, three problems emerge:

1. Customers drop off because they're unsure of the next step
2. Customers receive "checking in" emails from multiple departments simultaneously
3. Customers have to repeat the same information multiple times to different teams

Like our game of Pass the Message, each handoff risks losing crucial context about the customer's needs and progress, forcing them to start fresh with each new team they encounter.

4. Inconsistent messaging across the onboarding journey

The onboarding coordination problem also shows up as inconsistent messaging across the user journey. Your product's value proposition gets distorted as it travels across departments. Marketing materials emphasize certain benefits, Sales highlights different ones during demos, and Product tutorials focus on yet another set of features. Each team develops their own way of explaining the product:

- Marketing writes about "streamlining workflows."
- Sales pitches "operational efficiency."
- Product tutorials focus on "task automation."
- Support documentation uses "workflow automation."
- Customer Success talks about "process optimization."

Like our game of Pass the Message, each team translates your product's value into their own language, leaving customers confused about what they signed up for.

5. No one owns the end-to-end customer journey

In most organizations, each team owns its piece of the customer journey:

- Marketing owns awareness.
- Sales owns the buying process.
- Implementation owns technical setup.
- Customer Success owns customer relationships.
- Product owns the in-app experience.
- Support owns issue resolution.

But who owns the entire journey? Who ensures all these pieces fit together seamlessly? Often, the answer is "no one." Like our game of Pass the Message, without the party host who reveals the original message, no one ensures there's anyone to maintain the customer's experience and ensure it remains coherent from start to finish.

This lack of ownership of the end-to-end customer journey becomes even more complex when companies have both self-serve and high-touch onboarding processes. The product team might own the self-serve journey, while Customer Success owns the enterprise path. Without careful coordination, these parallel tracks can create inconsistent experiences and confusion about best practices. The solution isn't to force all customers down the same path but to keep both tracks aligned with your core onboarding principles.

THE IMPORTANCE OF CROSS-FUNCTIONAL ALIGNMENT

A fragmented onboarding experience is often the symptom of mis-alignment between teams. When departments operate in silos, the customer's journey becomes like our game of Pass the Message—each handoff risks losing crucial context, and the original vision gets distorted along the way.

Successful onboarding requires coordination across your entire organization. When teams align their efforts and communicate effectively, customers experience a smooth journey from signup to success—like passengers in a boat with a crew rowing in unison toward their destination. Without coordination, the boat spins in circles, and customers abandon ship.

To create a cohesive onboarding experience, three key elements must work together seamlessly, each owned and maintained by different teams. Let's examine each of these critical elements:

1. Positioning and Messaging

Marketing shapes how prospects understand your product before they even start. Strong positioning must answer four critical questions:

1. What is your product?
2. Who is it for?
3. What does it replace?
4. Why is it better?

If you don't nail these questions, users arrive with misaligned expectations, struggling to connect your product's value to their needs.

Outside of the product itself, I've found that positioning and messaging matter most for successful onboarding.

Strong positioning serves as the foundation for successful onboarding by creating a clear vision of success that makes each step feel purposeful. It shows users exactly when and how they'll use your product in their workflow while keeping them motivated during the initial setup by reinforcing the value they'll gain. Let's examine each of these aspects in detail:

1. Creates a Clear Vision of Success
Your positioning and messaging paint a picture of user success. They help users understand not just what your product does but how it will transform their current situation.

2. Sets Clear Expectations
Strong positioning and messaging show users exactly when and how they'll use your product in their workflow. As a result, they're more likely to know how to get started and what actions to take first.

3. Keep Users Motivated
The challenge with B2B onboarding is that users often need to invest significant time and effort before seeing value. Strong product messaging helps motivate new customers during this critical period by reinforcing the benefits of your product over their current approach. However, when Marketing isn't aligned with Sales and Customer Success, this messaging can backfire spectacularly. Marketing might promise "quick setup," while Implementation knows data migration takes weeks or highlights features that Sales knows aren't ready for enterprise use. These misalignments create a trust deficit from day one—customers feel oversold and underprepared, leading to frustration during onboarding and, ultimately, higher churn. Even worse, when Marketing emphasizes different value propositions than Sales or Customer Success, users enter onboarding with mismatched expectations about what success looks like, making it nearly impossible to deliver a satisfying experience.

2. Sales Pitch and Promises

The promises made during sales conversations become the expectations customers bring into onboarding. Sales demos shape the mental model of how your product works—which features matter, which workflows to follow, and what success looks like.

When sales teams promise "quick setup" or show sophisticated workflows that actually require significant configuration, they set expectations that can make or break the onboarding experience.

Sales conversations often include unofficial promises or "we can do that" moments that create hidden expectations. These might include:

- Timeline promises ("You'll be up and running in two weeks.")
- Feature availability ("Yes, that feature is coming next quarter.")
- Implementation support ("We'll help you migrate all your data.")
- Integration capabilities ("It works with all your existing tools.")

When these promises don't align with the actual onboarding experience or product capabilities, it creates what I call the "promise-delivery gap"—a major source of friction that can doom onboarding before it begins.

3. Product Education and Guidance

Product education bridges the gap between customer expectations and actual product mastery. In Chapter 3, we discussed the three pillars of successful onboarding: in-product guides, education content, and human touchpoints. These pillars must work in harmony to create a seamless experience.

- In-product guides show users what to do.
- Educational content explains why and how they should do it.
- Human touchpoints help when they get stuck.

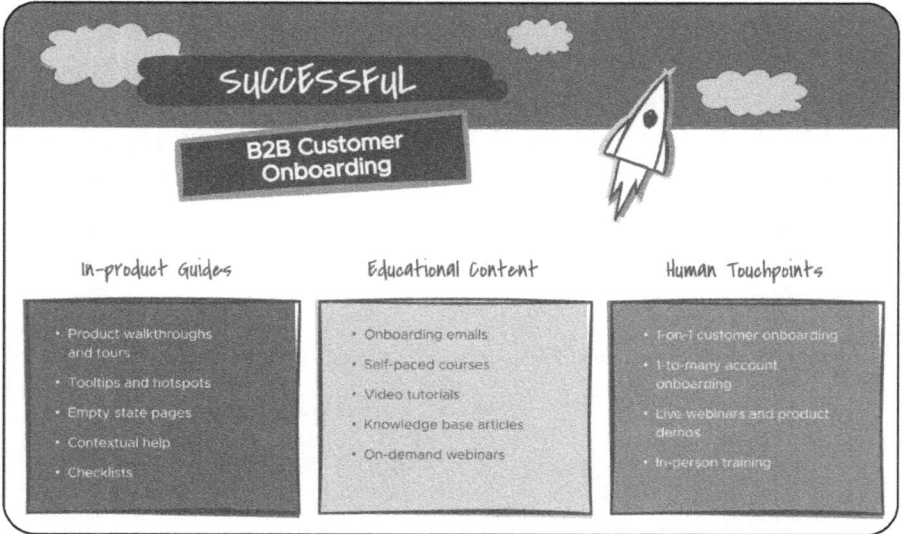

SUCCESSFUL

B2B Customer Onboarding

In-product Guides	Educational Content	Human Touchpoints
• Product walkthroughs and tours	• Onboarding emails	• 1-on-1 customer onboarding
• Tooltips and hotspots	• Self-paced courses	• 1-to-many account onboarding
• Empty state pages	• Video tutorials	• Live webinars and product demos
• Contextual help	• Knowledge base articles	• In-person training
• Checklists	• On-demand webinars	

Creating these pillars requires tight coordination between teams. Product needs Customer Success's insights about user struggles, Marketing needs Product's roadmap to create relevant content, and Support needs visibility into both to provide accurate assistance. When this coordination breaks down, the consequences are immediate: In-product guides use different terminology than Sales and Support materials, educational content becomes outdated or contradictory, and support representatives develop inconsistent solutions to common problems. What should be a clear path to success becomes a maze of conflicting information and mixed messages.

THE PATH FORWARD WITH A CROSS-FUNCTIONAL TEAM

But there's good news. These coordination challenges are solvable. In the next chapter, we'll explore how to build and structure an onboarding team that can create a seamless customer experience. Whether you're starting with informal champions or building a dedicated team, you'll learn how to get everyone rowing in the same direction.

Remember: The first step in the EUREKA Framework is establishing alignment because you can't solve cross-functional challenges with siloed solutions.

EUREKA ACTION ITEMS: ASSESS YOUR ORGANIZATION

Before diving into the next chapter, you can assess your organization's current state of cross-functional alignment. These action items will help you identify gaps in your onboarding coordination and prepare for improvements. Think of this as creating a baseline measurement—you can't improve what you haven't measured.

Review the five symptoms of an onboarding alignment problem discussed in this chapter and say if it's true or false at your company.

1. No clear definition of onboarding success.

- We don't have one clear activation definition that all teams agree on. True/False: _____

- Different departments use different metrics to measure onboarding success. True/False: _____

- We haven't documented what "good onboarding" looks like. True/False: _____

2. No single source of truth for customer data.

- Customer data is scattered across multiple tools without synchronization. True/False: _____

- We can't easily track a customer's complete onboarding progress. True/False: _____

3. Unclear onboarding handoffs between teams.

- There's no standard process for transferring customers between teams. True/False: _____

- Customers often have to repeat information to different teams. True/False: _____

- Important context gets lost during team transitions. True/False: _____

4. Inconsistent messaging across the onboarding journey.

- Our website promises differ from what Sales tells customers. True/False: _____

- Product tutorials use terminology different from our marketing materials. True/False: _____

- Different teams describe the same features using inconsistent terms. True/False: _____

5. No one owns the end-to-end customer journey.

- There's no single person or team responsible for the entire onboarding experience. True/False: _____

- Different teams make onboarding decisions without coordinating with each other. True/False: _____

- No one regularly reviews the complete customer journey from start to finish. True/False: _____

5

The Onboarding Team and Structure

～～～

Individual commitment to a group effort—that is what makes a team work, a company work, a society work, a civilization work.

Vince Lombardi

～～～

In college, I was part of a dragon boat racing team. Unlike traditional rowing, where athletes face backward, dragon boat racers face forward—twenty people in a long boat, each with a paddle, all looking toward their shared destination as they race across the water. The boat with the fastest time wins.

What I learned quickly was that raw power alone doesn't win races. Sure, you need strength to dig deep with each stroke, but that's not enough. The winning teams are the ones who row in perfect synchronization while maintaining a shared vision of their destination. This forward-facing alignment mirrors what we discussed in the previous chapter: Teams need to not only work in sync but also share a clear vision of where they're heading. Every paddler needs to enter and exit the water at exactly the same moment, working as one unit. A single person out of sync can actually slow the boat down, no matter how strong they are.

The same principle applies to fixing your onboarding experience. Individual effort, no matter how heroic, isn't enough. You need a coordinated team effort, with everyone moving in the same direction at the same pace. Marketing, Product, Customer Success, and Sales must work in harmony, like paddlers in sync, to create momentum and drive results.

In the previous chapter, we explored why cross-functional coordination is crucial for successful onboarding. We looked at how misalignment between departments creates a "broken telephone" effect, where customer expectations and product messaging get distorted as they pass between teams. We also identified five key symptoms of poor onboarding coordination: unclear success definitions, fragmented customer data, messy handoffs, inconsistent messaging, and lack of end-to-end ownership.

In this chapter, we'll explore how to build and structure your onboarding team. Whether you're starting with an informal group of champions or building a dedicated activation team, you'll learn how to get everyone rowing in the same direction.

KEY ROLES AND RESPONSIBILITIES

The process to improve user onboarding works best when it's a team effort, ideally from across different functions within a company. While team structures and sizes vary widely based on company needs and resources, the core mission remains the same: supporting new users toward success. Let's examine the key roles that typically contribute to this goal.

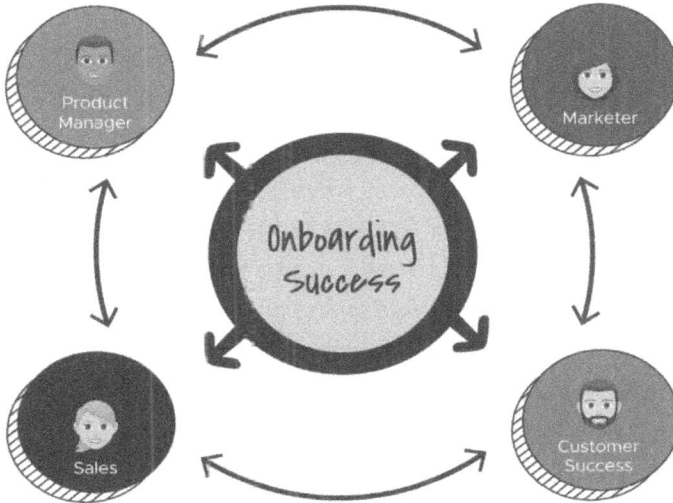

Product Managers

Product managers typically orchestrate the in-app user onboarding experience with designers and engineers, from the signup to the first-use workflow. They also oversee the implementation of any triggers inside your product. This includes (but is not limited to) progress bars, product tours, and checklists that help guide new users to their desired outcomes.

Product managers are also responsible for measuring and improving key activation metrics. They analyze user behavior data to identify where users get stuck, conduct user research to understand pain points, and run experiments to optimize the onboarding flow. By working closely with analytics teams, they can track important metrics like time-to-value, feature adoption rates, and user progression through key milestones.

Marketers

Marketers communicate a product's value and educate new users to become effective, regular users. This could include crafting and managing life cycle email campaigns, templates, case studies, and helpful tips—each designed to draw the user's attention to the product's core actions. By using out-of-product guides such as onboarding emails, browser notifications, SMS messages, direct mailers, or retargeting ads (normally the job of a marketer), you can remind users how the product improves their lives.

Furthermore, great marketing builds trust and creates a personal, emotional connection. This hits every touchpoint: from ads, blog posts, and landing pages to onboarding copy, in-app messages, and emails. A marketing team needs to create a cohesive and consistent content strategy that amplifies the need for a product or the pain of their situation to help them overcome their anxieties and objections.

Customer Success

Those responsible for customer success and happiness should lend a huge hand in the onboarding process. Understanding a user's wants and needs is crucial during onboarding. Customer success teams are usually equipped with the empathy and product know-how necessary to show users immediate value.

When users stumble during the user onboarding process, it's up to customer success to reach out and measure how they feel as they progress. Since they're the first point of contact for issues and problems new users face, they can provide invaluable user feedback for the business. Every support question and survey response should be recorded and relayed to where it's needed to enhance the user onboarding experience.

Sales

With hybrid or sales-assisted onboarding, the sales team reaches out to potential customers to ensure they're receiving a lot of value from a product. Salespeople tend to take special care of individual users while laying out expectations for further features of the product to motivate them to build a deeper, more frequent habit with a service, such as premium features or subscription benefits.

Using product engagement data from the product and marketing teams, sales teams often use demos to provide customized walk-throughs of the product. This is how two teams can work hand in hand: The sales team can build on the relationships first initiated by the marketing team.

Now that we understand the key players involved in onboarding, the question becomes: How do you organize these roles into an effective team? The answer depends on your company's size, resources, and level of commitment to improving the onboarding experience.

THE THREE LEVELS OF ONBOARDING TEAM STRUCTURE

Based on my experience with B2B companies, there are three distinct approaches to structuring an onboarding team. Each level represents a different degree of organizational commitment and resources, with its own advantages and tradeoffs. Let's explore which structure might work best for your organization.

Level 1: Informal Structure

When you're just beginning to improve your onboarding process, you might not have a formal mandate from leadership to create a dedicated team. Rather than waiting for official approval and structure, you can start building momentum by identifying key champions across departments.

The key is finding allies who understand the importance of effective onboarding and are willing to collaborate informally. Look for:

- Product managers who consistently bring up activation data in meetings or have already started experimenting with onboarding improvements.

- Customer success managers who've created their own onboarding checklists or documentation to help their customers succeed.

- Sales team members who take detailed notes about why trials don't convert or actively seek feedback from churned customers.

- Marketing team members who've identified gaps between marketing promises and the actual customer experience.

Start by having informal conversations with these potential champions. Share the exercises and activities from Steps 2 to 5 of the EUREKA Framework (which we'll cover in upcoming chapters) to gather their perspectives on what successful onboarding looks like. Ask questions like:

- What does onboarding success mean for your role?
- What are the biggest onboarding pain points you see?
- How could better onboarding impact your team's goals?

The goal is to create a collaborative environment where different perspectives on onboarding can be freely shared. By bringing together these informal champions, you can start identifying common pain points and opportunities for improvement across departments. This grassroots approach may not have official authority, but it can create meaningful improvements through shared understanding and voluntary cooperation. It can also set you up for getting formal buy-in later when you can demonstrate early wins and alignment.

Let's examine the advantages and limitations of this informal approach:

Pros:

- Immediate start without waiting for formal approval

- Get early wins to build the case for a formal structure

- Room for experimentation before scaling

- Organic buy-in across departments

Cons:

- Heavy coordination burden on champion

- Limited resources and official support

- Difficulty driving systematic changes

- Slower progress without dedicated resources

- Risk of stalling if key champions leave

Even with these limitations, starting small is better than not starting at all. Document your learnings, measure improvements where possible, and use these early results to support a more formal structure when the time comes.

Level 2: Decentralized Onboarding Squad

In this level, a decentralized onboarding team—also sometimes called an embedded team or a tiger team—consists of team members pulled from other departments within an organization. Unlike Level 1's grassroots effort, this model has official leadership support and dedicated meeting time. These champions still maintain their primary roles, but they now have formal authority to coordinate onboarding initiatives and implement changes within their departments.

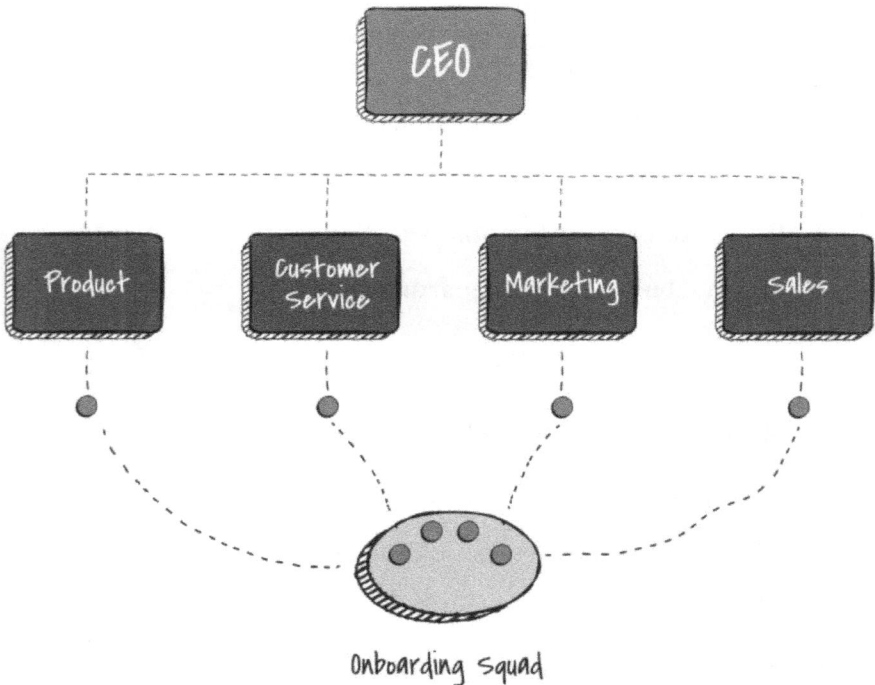

Onboarding Squad

During my time at Appcues, I was part of an activation tiger team focused on improving our trial-to-paid conversion rate. Our team included champions from Product, Marketing, Customer Success, and Sales, each bringing unique perspectives and capabilities to improve our activation rates.

We met bi-weekly to:

- Review activation metrics and identify friction points
- Brainstorm experiment ideas across touchpoints
- Share results from previous experiments
- Coordinate upcoming initiatives

One such effort resulted in double-digit gains for their activation rate (from less than 2 percent to 25 percent) by breaking down their onboarding into bite-sized steps. Through coordinated efforts across departments, we segmented and personalized the initial product setup (Product), created targeted onboarding emails (Marketing), and developed better video training resources (Customer Success).

Let's examine the advantages and limitations of this decentralized approach:

Pros:

- Clear ownership and accountability across touchpoints
- Authority to implement changes within departments
- Regular coordination of cross-functional efforts
- Direct access to departmental resources
- Strong connection to team priorities and goals

Cons:

- Split focus between onboarding and primary roles
- Risk of communication silos
- Limited resources for major initiatives
- Dependence on individual champion influence

The decentralized squad model works well for companies that want to improve onboarding without creating a dedicated, full-time team. It provides structure and accountability through a formal cross-functional team while allowing members to maintain their primary departmental roles.

Level 3: Centralized Activation Growth Team

The highest level of organizational commitment to improving activation is creating a dedicated, centralized team focused solely on onboarding and activation metrics. Unlike the decentralized model, where team members split their time, a centralized activation team consists of individuals who dedicate 100 percent of their time to improving key activation metrics like monthly active users, time-to-value, and trial-to-paid conversion rates.

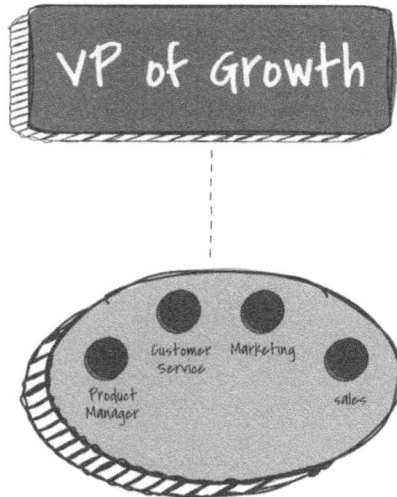

This team typically reports directly to a senior leader (VP of Growth, Chief Product Officer, VP of Customer Experience, etc.) and has the authority to run experiments and make changes across the entire customer journey. A typical centralized activation team might include:

- Product Manager focused on product onboarding and activation
- Growth Engineer(s) for quick experimentation
- Data Analyst for measuring impact
- Designer specialized in user experience
- Content Strategist for educational materials
- Customer Success for onboarding strategy and support

The team operates with significant autonomy and a clear mandate: Improve activation metrics through rapid experimentation and iteration. They own the entire activation journey, from initial signup through to active product usage, and can quickly implement changes without navigating complex cross-departmental approvals.

Let's examine the advantages and limitations of this centralized approach:

Pros:

- Maximum speed and experiment velocity
- Full ownership and dedicated resources
- Direct access to leadership decisions
- Clear metrics and accountability
- Deep activation expertise development

Cons:

- High cost of dedicated headcount
- Potential disconnect from departments

- Friction with existing teams
- Heavy organizational investment needed
- Limited broader business context

This model works best for larger organizations where activation improvements can justify dedicated resources. Companies like Slack, Dropbox, and HubSpot have successfully used centralized growth teams to optimize their activation metrics and drive sustainable growth.

THE RISE OF DIGITAL AND SCALED CUSTOMER SUCCESS TEAMS

An emerging trend among growing B2B companies is consolidating onboarding ownership under Digital Customer Success or Scaled Customer Success teams. These teams typically sit within the Customer Experience organization and are responsible for creating scalable onboarding experiences across multiple touchpoints.

For example, at Fullstory, their Digital Customer Success team owns the entire onboarding journey—from welcome emails to in-product guidance and educational content. As Chrissy Quinones, Digital Customer Success Program Manager at Fullstory, explains:

> **"Our team falls across the entire customer journey, with onboarding being a critical piece that we own. While we have CSMs working directly with accounts and offering paid onboarding services, we focus on creating scalable experiences for all users—whether they're part of the initial purchase or joining an existing account later."**

Their four-person Digital Customer Success team is led by a Senior Manager who reports to a VP of Digital Customer Success and Support. They work alongside but separate from the traditional CSM organization. This structure allows them to move quickly and experiment with different approaches to activation while maintaining a consistent experience across all customer touchpoints.

The advantage of this structure is that it provides clear ownership of the onboarding experience while maintaining the agility to experiment and iterate quickly. These teams can coordinate with Product, Marketing, and CSM teams while maintaining autonomous execution

While this structure may not make sense for every organization today, I believe it represents the future of B2B onboarding. As companies scale, the focus shifts from managing customer accounts to engaging individual users. Companies can better drive adoption across their entire customer base by creating dedicated teams that can scale user success through digital touchpoints. **After all, customer success starts with user success.**

GETTING STARTED: THE PATH OF LEAST RESISTANCE

Your onboarding team structure is not set in stone. As Adam Fishman, former Head of Growth at Lyft, explains in a Reforge article:

> **"One of these organizational structures is not what the company needs forever. Companies will move through these stages and will kind of move back and forth depending on some different variables around velocity, around culture, around whether or not you're the bottleneck to progress."**

Fishman provides the example of Elena Verna's experience at SurveyMonkey. As the former SVP of Growth, she regularly adjusted the growth team structure based on the company's changing needs. While she initially led a centralized team to drive rapid experimentation and improvements, they later transitioned to a more decentralized structure when the core teams developed stronger growth capabilities and the organization's priorities shifted.

What matters most is choosing a structure that:

- Aligns with your current activation goals
- Fits your available resources
- Matches your organization's culture
- Enables the speed of execution you need
- Supports effective cross-functional collaboration

A mentor once shared valuable advice that applies to any team structure: "Move quickly and touch lightly." Look for the path of least resistance and make progress in short steps.

Ask yourself: "*What is the smallest, easiest step I can take that moves me in the right direction?*" If you're already working on fixing the onboarding for your product, you can start at level 1 by identifying and connecting with potential champions across departments who own or manage different components of your onboarding experience.

As your company grows and activation becomes a strategic priority, consider moving to level 2's decentralized squad. This works well when you have established departments but need better coordination. Level 3's dedicated team makes sense for larger organizations where activation metrics directly impact revenue growth and can justify dedicated headcount.

Remember, there's no one-size-fits-all approach. Your team structure should evolve with your company's growth, resources, and activation challenges.

EUREKA ACTION ITEM: IDENTIFY YOUR ONBOARDING TEAM

Decide who should be involved and what their primary responsibilities are. If you're wondering who should be in your onboarding team, ask yourself which team or person is responsible right now for:

- Product positioning and messaging
- User experience and workflows
- Onboarding communications
- Customer support and success
- Sales conversion

To help you map out your onboarding team structure and responsibilities, I've created a free worksheet that you can download at *eurekabonus.com*.

Onboarding Team Worksheet

	Product	Marketing	Customer Success	Sales
Name				
Role				
What does "user onboarding" mean to you?				
How do we know we've successfully onboarded a new user?				

In the next chapter, I'll share a 60 workshop activity that you can facilitate to align these champions around shared goals and onboarding opportunities.

6

The Onboarding Sailboat Exercise

～～～

**To reach a port we must set sail – Sail,
not tie at anchor. Sail, not drift.**

Franklin D. Roosevelt

～～～

Building and maintaining an effective onboarding team comes with several challenges. These include managing competing priorities, maintaining momentum, and coordinating efforts across departments. I'll cover these more in-depth in Step 5 of the EUREKA Framework.

One of my strategies to address these challenges is through facilitated onboarding workshops. These structured sessions bring together stakeholders from different departments to align on goals and surface challenges and create shared understanding around the onboarding experience.

These workshops serve multiple purposes:

- Surface different perspectives on what successful onboarding looks like
- Highlight misalignments in priorities and expectations
- Create shared ownership of onboarding improvements
- Build relationships across departments
- Generate actionable next steps

When I work with companies to improve their onboarding, I start with collaborative exercises that get everyone involved and think critically about the current experience. These activities help teams move from abstract discussions to concrete actions.

In the next chapter, we'll explore how to get your entire team aligned on what's working and what isn't in your onboarding. The Onboarding Sailboat Exercise, adapted from agile retrospectives, provides a creative and engaging way to visualize these challenges and get everyone on the same page.

OBJECTIVE

The Onboarding Sailboat Exercise serves multiple objectives:

- Create alignment among team members about current onboarding challenges
- Identify and prioritize obstacles in your onboarding process
- Surface positive elements that are already working well

By the end of this exercise, your team will have a prioritized list of challenges and a clear understanding of what's working, setting the foundation for targeted improvements in your onboarding process.

THE PARTICIPANTS

This exercise works best with the cross-functional team you assembled in Chapter 5. Include team members from key functions responsible for onboarding:

- Product Management: to represent product vision and roadmap
- Design: to speak to user experience and interface
- Engineering: to provide technical context
- Customer Success: to share front-line user feedback
- Sales: to bring insights from prospect conversations
- Marketing: to align messaging with user needs

Ideally, choose participants who work directly with onboarding and can contribute meaningful insights. Remember, you want quality of discussion over quantity of participants.

Keep your group small—no more than eight participants. Too many voices can make the exercise unwieldy and less effective.

THE SETUP

The exercise uses a powerful metaphor to help teams visualize factors affecting their user onboarding:

- The wind in the sails represents forces pushing users forward in their onboarding journey.
- The anchor and rocks below represent obstacles holding users back.
- The island in the distance represents successful user onboarding.

This visual metaphor makes it easy for team members to contribute their perspectives and create a shared understanding of your onboarding challenges.

Draw the sailboat scene on a whiteboard (have some fun with it!), being sure to include the following visual elements:

- The wind in the boat's sails
- The anchor dragging behind it

Time needed: 30–60 minutes

Materials needed:

- Whiteboard or digital whiteboard
- Square sticky notes (two different colors)
- Sharpies or markers
- Voting dots (8 per participant)

You can also use a template I've created for FigJam and Miro. You can access it at *eurekabonus.com.*

FACILITATION TIPS

To get the most out of the Onboarding Sailboat Exercise, consider these facilitation best practices:

- Send participants a brief overview beforehand (template available at *eurekabonus.com*)

- Play background music during silent activities to reduce awkwardness

- Keep the group small (under eight people) with representatives from each key function

- Stick to timeboxes and have all materials ready before starting

- Stay neutral and focused on user experience if discussions become heated

THE ACTIVITY

Part 1: What's Moving New Users Forward

1. Give each team member a block of square sticky notes and a Sharpie.

2. Ask them to silently write down as many things as possible that are moving new users forward to achieving their goals. Prompt them to think about what's working well in your current onboarding process. Give each participant 3 to 5 minutes to write them down silently. They should write simple statements, one per sticky note.

Examples might include:

- ○ Providing resources through our product academy
- ○ Sales reaching out to customers if they have problems
  - • Clear getting started checklist
  - • Helpful onboarding emails
  - • Responsive support team

3. Once the time is up, you, as the facilitator, will ask each participant, one by one, to stick their sticky notes to the top part of the Sailboat.

4. Ask each participant to call out one thing that they put down.

This step creates a positive atmosphere by starting with what's working well. When team members hear about the good things happening in their onboarding process, they build confidence and create an encouraging foundation before diving into challenges.

Part 2: What's Holding New Users Back?

1. Ask them to silently write down as many things as possible that are holding back new users from achieving their goals. What's preventing us from successfully onboarding more customers? Tell participants that their stickies will be anonymous this time, so they should write whatever they like.

2. After 5 to 10 minutes, ask everybody to stick their sticky notes to the bottom of the Sailboat. They should do this fast and randomly without discussion. If there isn't enough space underneath the sailboat, simply have them spread out on the whiteboard.

3. Ask people to help you organize sticky notes on the same challenges. For example, if one sticky says, "Users feel lost without any guidance," and another says, "No clear path for new users," group them together. Look for patterns and themes in the challenges identified.

4. Once grouped, give each cluster a descriptive label that captures the theme.

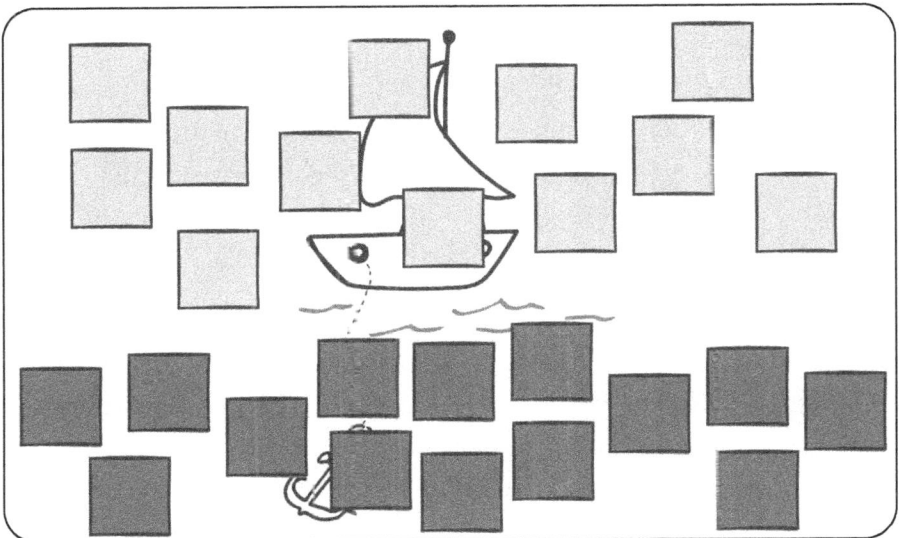

Part 3: What's Causing the Most Troubles?

In this part, we'll use dot voting to prioritize the challenges we've identified. Dot voting (also known as dotmocracy or multi-voting) is a simple prioritization technique in which participants place colored dot stickers on items they think are most important. It's a quick and visual way to see what the group considers the highest priority.

1. Give each participant 8 voting dots. You can do something similar with FigJam or Miro. See how at eurekabook.com/resources.

2. Set the timer to 8 minutes and tell them to vote on what they consider to be the biggest issues holding the team back when it comes to the problem we're tackling. Remind them of the voting rules:

    o They can vote on their own stickies.
    o They can put as many votes on one sticky as they like.
    o They need to use all their votes.
    o No talking is allowed.

3. Once the 8 minutes are up, tell everybody to sit down and create a vertical stack of the stickies in order of most votes to least; ignore anything with 2 votes or fewer.

4. Discuss with the team if these problems resonate with them. Are there things on the board that surprised them?

THE OUTCOME

This exercise isn't about finding solutions—yet. Its primary purpose is to create alignment around your team's onboarding challenges. You might be surprised to discover that what seems obvious to one team member isn't apparent to others. By the end of this exercise, you'll have:

- A clear picture of what's working well in your onboarding
- A prioritized list of challenges to address
- Team alignment on the most pressing issues

With these insights, you'll be ready to tackle these challenges using the remaining steps of the EUREKA Framework in the following chapters.

EUREKA ACTION ITEMS: THE ONBOARDING SAILBOAT EXERCISE

Ready to run your own Onboarding Sailboat Exercise? Here's your action plan:

1. Gather your cross-functional team (aim for four to eight people)
2. Download the exercise template from *eurekabonus.com*
3. Schedule a sixty-minute session
4. Follow the three-part structure:
   - Identify what's moving users forward
   - Uncover what's holding users back
   - Vote on the biggest challenges

Remember, the goal is to align your team with your current onboarding challenges. Don't worry about solutions yet—we'll get there in the upcoming chapters.

In Chapter 7, we'll move to the next step in the EUREKA Framework: Understanding User Success. We'll take the challenges you've identified in the Sailboat Exercise and learn how to deeply understand what's causing them through user research and data analysis.

STEP 2

Understand User Success

Establish a Team	Understand Success	Reverse Journey Map	Keep Users Engaged	Apply Analyze and Repeat

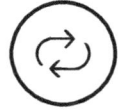

7

The Three Dimensions of B2B User Success

~~~

**If you don't know where you're going,
you'll end up someplace else.**

Yogi Berra

~~~

Imagine spending months building a bridge, only to realize you've connected it to the wrong destination. The engineering is flawless, the materials are top-quality, and the construction is solid—but none of that matters because people need to cross to a different place entirely.

Many companies make this exact mistake when it comes to product onboarding. They invest heavily in product tours, email sequences, and documentation without first understanding where their users need to go.

As we discussed in Chapter 1, onboarding is the bridge between users' current situation and their desired outcome, but you need to know the right destination before you start building. That's why defining user success is the crucial second step of the EUREKA Framework. Without this foundation, even flawlessly executed onboarding tactics will fail.

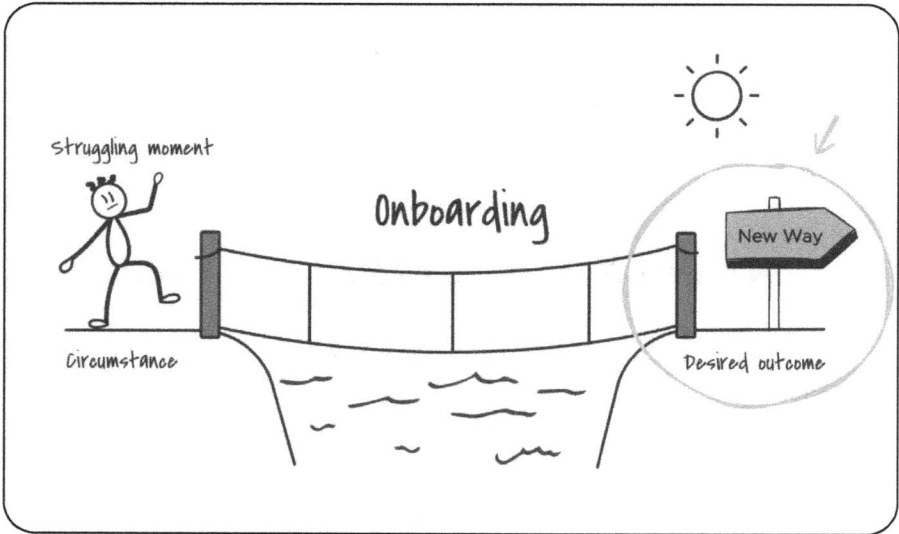

This chapter explores how to define user success for B2B products. We'll start with the Jobs-to-Be-Done (JTBD) framework to understand the core transformation users seek. Then, we'll examine how this manifests across three dimensions: the functional, emotional, and social aspects of success. Finally, we'll expand this framework to address the complexity of B2B onboarding, where multiple stakeholders, user roles, and use cases each require their own JTBD statements and success criteria. But first, let's understand onboarding's true purpose: enabling transformation, not just teaching features.

ONBOARDING'S ULTIMATE GOAL: LEVEL UP THE USERS

At its core, customers don't buy B2B software to learn new features—they buy it to become better at their jobs. As we saw in Chapter 2, people naturally resist change. They'll only adopt new tools when they see a path to becoming more productive, more effective, or more successful in their careers.

Samuel Hulick from UserOnboard captures this perfectly with his Super Mario analogy: Players don't get excited about the fire flower because it's pretty—they get excited because it transforms Mario into a more powerful version of himself. Your product is that power-up for your users.

This isn't what your business makes

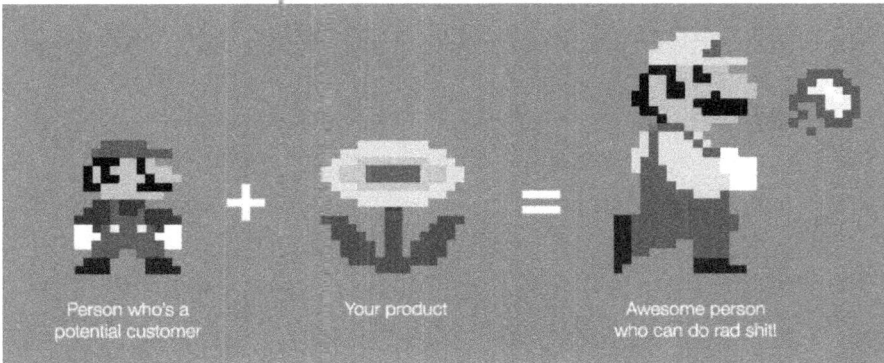

Person who's a potential customer + Your product = Awesome person who can do rad shit!

@UserOnboard

This is

This mindset shift changes everything: **Your product isn't just a tool but a catalyst for transformation.** Asana transforms scattered managers into organized leaders. Salesforce turns sales reps into data-driven professionals. Zoom enables teams to become expert remote collaborators.

Kathy Sierra, author of *Badass: Making Users Awesome,* says it best: "Upgrade your user, not your product. Don't build better cameras—build better photographers."

So before we explore how to define and measure user success, we need to answer one crucial question: What transformation are your users really seeking? That's where the Jobs-to-Be-Done Framework can help.

THE JOBS-TO-BE-DONE FRAMEWORK

The concept of user transformation aligns perfectly with the Jobs-to-Be-Done (JTBD) framework developed by Harvard Business School professor Clay Christensen. In his groundbreaking book *Competing Against Luck*, Christensen explains that people don't simply buy products—they "hire" them to make progress in their lives.

Think about it: When a B2B customer purchases your product, they're not just acquiring software. They're hiring your solution to help them transform their current situation into a better one. For example:

- A marketing team doesn't just buy an email marketing platform—they hire it to build stronger customer relationships.

- A finance department doesn't just buy accounting software—they hire it to improve the accuracy and efficiency of company finances.

- A product team doesn't just buy analytics tools—they hire them to become more data-driven in their decision-making.

That's where JTBD statements can help you define the transformation your product enables. It follows a simple format:

"When I [situation], I want to [motivation], so I can [expected outcome]."

For example, a JTBD statement for a project management tool might be: "When managing multiple team projects, I want to keep everything organized and visible, so I can deliver results on time and keep stakeholders informed."

There are three key observations about the JTBD statement that are particularly relevant for B2B onboarding:

1. JTBD statements are solution-agnostic

Notice how JTBD statements never mention specific tools or solutions—they purely describe the situation, motivation, and desired outcome. When a team leader says, "When managing a growing team (situation), I want to keep everyone aligned on priorities (motivation), so we can hit our targets consistently (outcome)," they're describing their needs independent of any solution. They might solve this through project management software, spreadsheets, regular meetings, or other approaches. Understanding this helps you focus onboarding on the transformation users seek rather than just teaching features.

2. Circumstances create "job openings"

People don't wake up randomly deciding to buy software. Instead, specific circumstances create metaphorical "job openings" that need to be filled—perhaps by your solution. When a team grows too large for spreadsheets to manage or a company lands a big client requiring better reporting, these situations trigger the search for solutions. It's like a job posting going live: Suddenly, there's a clear need that must be addressed. Understanding these triggering circumstances provides crucial context for your onboarding experience, as they reveal why users are seeking change right now.

3. Expected outcomes drive adoption

The desired outcome in a JTBD statement reveals what truly motivates users to change their behavior. When users say they want to "keep everything organized," the real expected outcome might be "looking competent in front of my boss" or "reducing stress from missed deadlines." Understanding these deeper outcomes helps you design onboarding experiences that connect product features to meaningful user goals.

These deeper outcomes reveal an important truth about user success: It's multidimensional. When we look closely at the expected outcomes in JTBD statements, we see that users aren't just seeking functional benefits to overcome technical barriers—they're looking for emotional satisfaction that addresses their anxieties and social validation that helps them drive organizational change. This understanding leads us to the three dimensions of user success that mirror those friction levels.

THE THREE DIMENSIONS OF USER SUCCESS

User success in B2B products directly mirrors the hierarchy of friction we explored in Chapter 2. Just as users face functional, emotional, and social barriers, they need corresponding dimensions of success to create lasting transformation.

Emotional Friction
Fears, doubts, and anxieties about the change of adopting a new product.

Social Friction
Barriers involving people, processes, and politics within the user's org.

Functional Friction
UX/UI product issues that hinder users from using key features and workflows.

Each dimension connects to a distinct sensory experience:

- Functional success is what users can **see** themselves accomplish.

- Emotional success is what users **feel** when using your product.

- Social success is what users **hear** others say about them.

Let's explore each dimension, starting with the most misunderstood:

1. Functional Success (See)

Functional success is about the transformation your product enables—the new superpower users gain that they didn't have before. It's not about mastering product features; it's about the broader transformation of how they work. Users don't just learn to use new tools—they evolve into more effective professionals who can accomplish things that were impossible before.

Think about Figma. Functional success isn't just about creating designs or sharing files. It's about enabling designers to build and iterate faster than ever before. It's about giving them the superpower to collaborate in real time, eliminating the endless back and forth of traditional design workflows.

A common mistake is being too shallow when defining your product's functional success. Many teams stop at surface-level outcomes, missing the deeper transformation users seek. One way to dig deeper is to ask, "So what?" repeatedly until you reach the core outcome.

Let's apply this to a project management software like Asana:

- "Our product helps teams track their tasks." ***So what?***

- "They can see who's working on what." ***So what?***

- "They can identify and remove bottlenecks quickly." ***So what?***

- "They can ensure every project stays on schedule." ***So what?***

- "They can consistently deliver successful projects while maintaining team morale."

The "So what?" exercise reveals that Asana's true functional success isn't about task management or project tracking—it's about transforming project managers from overwhelmed coordinators into confident orchestrators who can predictably deliver successful outcomes. That's the superpower users are really hiring Asana to gain.

To identify your product's true functional success, ask yourself:

- What are things users ***see*** themselves doing now that they couldn't do before?

- What desired outcome do they now ***see*** realized?

- What needs do they ***see*** now satisfied?

2. Emotional Success (Feel)

While functional success focuses on what users can do, emotional success centers on how users feel when using your product. Every B2B purchase, despite being "rational" on the surface, carries significant emotional weight. B2B users aren't just looking to accomplish tasks— they want to feel certain ways and avoid feeling others.

PRO TIP: WRITE FROM YOUR USERS' PERSPECTIVE

When answering these questions and the ones for emotional and social success below, write from your ideal user's first-person perspective. Instead of writing, Users can deliver projects on time, write, I see myself delivering projects consistently on time and under budget.

This shift in perspective helps you better empathize with your users and uncover deeper insights about their desired transformation.

Take our project management software example. A project manager's emotional success isn't just about delivering projects on time. It's about:

Avoiding negative emotions like:

- Fear of missing critical deadlines
- Anxiety about losing track of important details
- Frustration from constant status update meetings
- Embarrassment when caught off-guard by stakeholders
- Overwhelm from juggling multiple projects
- Stress from unexpected bottlenecks

And achieving positive emotions like:

- Confidence in knowing everything is on track

- Pride in running efficient, productive meetings

- Relief from having a single source of truth

- Satisfaction when projects complete smoothly

- Peace of mind from having clear visibility

- Recognition for being highly organized

A project manager who delivers on time but feels constantly stressed hasn't truly succeeded with your product. That's why onboarding must address both functional and emotional outcomes—helping users feel confident and in control, not just teaching them features.

For your product, ask yourself:

- What are the positive emotions users *feel* after achieving their desired outcome?

- What are the negative emotions they avoid *feeling* after achieving their desired outcome?

Understanding these emotional drivers will help you design an onboarding experience that doesn't just teach features but builds confidence and reduces anxiety throughout the journey.

3. Social Success (Hear)

The final component of user success is social—how users want to be perceived by others. B2B software, despite its business focus, is fundamentally about human connections. The tools we choose at work shape how our colleagues, managers, and customers perceive our capabilities.

Let's continue with our project management software example. A project manager's social success isn't just about delivering projects—it's about how they're perceived within their organization. They want to hear:

- Their boss says: "I can always count on your team to hit deadlines."

- Their company's executives say: "Your quarterly project reviews are so clear and data-driven. Other departments should follow your approach."

- Colleagues say: "Your planning makes it so much easier to stay organized."

- Other project managers say: "Can you show me how you set up your workflows? Your system is impressive."

- Stakeholders say: "I always know exactly where things stand with your projects. No surprises."

For your users, social success translates directly into career advancement. By championing a successful solution, they're not just solving a business problem but positioning themselves for a promotion, raise, or company recognition.

Social success is crucial for B2B products because growth depends on internal champions. Your users need to become advocates who can convince other teams to adopt your solution. They might achieve functional success (like delivering projects on time) and emotional success (feeling confident), but if they can't persuade others to embrace the tool, both their career growth and your product's expansion hit a ceiling.

For your product, ask yourself what you want your users to **hear** their colleagues, managers, or customers say about them.

Understanding these social aspirations will help you design onboarding experiences that enable users to succeed and showcase that success to others.

THE COMPLEXITY OF B2B ONBOARDING SUCCESS

B2B onboarding is inherently more complex because products typically serve multiple users within an organization—each with its own JTBD statements and dimensions of user success. A project management tool might be needed to satisfy executives who are tracking ROI, managers coordinating teams, and individual contributors who are completing tasks. Each user type brings its own JTBD statement and requires different functional, emotional, and social wins to feel successful. Let's examine three key ways this complexity manifests in B2B products:

1. Buyers vs. End Users

In B2B software, buyers and end users often have fundamentally different JTBD statements and success criteria. Take a project management tool, for example:

Buyers (Executives):
JTBD: "When managing multiple teams, I want to standardize project delivery processes so I can improve operational efficiency and reduce project delays."

Success Dimensions:

- Functional: **See** improved on-time delivery rates and resource utilization
- Emotional: **Feel** confident about project status and team productivity
- Social: **Hear** board members praise operational improvements

End Users (Project Managers):

JTBD: "When coordinating daily tasks, I want to keep everything organized and visible so I can deliver results without constant status meetings."

Success Dimensions:

- Functional: **See** all project tasks and deadlines in one place
- Emotional: **Feel** less stressed about missing important details
- Social: **Hear** team members say how much easier collaboration has become

Your onboarding must satisfy both perspectives: helping buyers track organizational wins while enabling end users to experience daily workflow improvements. Missing either group's success criteria risks poor adoption or eventual churn.

2. User Roles

Even within the same team, different roles often seek different transformations. Take a design tool like Figma: Designers want to create and iterate on designs faster, product managers want to give feedback more efficiently, and developers want to extract accurate specs more easily. Each role's JTBD reflects its unique goals—from "When designing features, I want to quickly test variations so I can find the best solution" to "When implementing designs, I want to access exact measurements and styles so I can build pixel-perfect features."

3. Use Cases

Success definitions also vary across different use cases. Each scenario requires its own JTBD statement to clarify the specific transformation users seek, helping you create targeted onboarding experiences that address each group's unique needs.

For instance, a project management tool serves different needs for IT teams running deployments versus marketing teams coordinating campaigns. The Marketing team's JTBD might be, "When planning campaigns, I want to coordinate multiple deadlines and deliverables so I can launch on time without dropping any balls." Meanwhile, an IT team might be, "When managing deployments, I want to track dependencies and risks so I can release new features without disrupting users."

BRINGING IT ALL TOGETHER: USER SUCCESS IN B2B PRODUCTS

In this chapter, we explored how to define success for B2B products using the Jobs-to-Be-Done framework. We learned that true user success isn't just about teaching features—it's about enabling transformation across three key dimensions:

- Functional success: What users can **see** themselves accomplish

- Emotional success: What users **feel** when using your product

- Social success: What users **hear** others say about them

We then examined how these dimensions become more complex in B2B contexts, where success must be defined differently for:

- Buyers versus end users, each with their own JTBD statements

- Different user roles within the same team

- Various use cases across departments

Understanding these multiple layers of success is crucial for creating onboarding experiences that drive meaningful adoption and transformation across entire organizations.

EUREKA ACTION ITEMS: DEFINE USER SUCCESS

Understanding these components and variations is crucial for creating onboarding experiences that drive meaningful adoption and transformation.

Take some time to define the user success for your product's primary use case:

1. Create Your JTBD Statement

- Who is your primary user?

- What situation triggers their need?

- What motivation drives them?

- What outcome do they seek? Write it out: "When [situation], I want to [motivation], so I can [expected outcome]."

2. Define Functional Success (See)

- What new capabilities will users gain?
- What can they do now that was impossible before?

3. Define Emotional Success (Feel)

- What negative emotions will they avoid?
- What positive emotions will they experience?

4. Define Social Success (Hear)

- What will their boss or manager say about them?
- What will their peers or customers notice?

Start with your primary user type, then repeat these exercises for other key stakeholders (like buyers) and different use cases. This will help you build a comprehensive picture of success across your entire B2B customer base.

Later in Chapter 9, I'll share the User Success Canvas—a collaborative workshop exercise that helps your onboarding team align on these three dimensions of success. You'll learn how to map success criteria across different user roles and use cases, creating a foundation for targeted onboarding experiences.

Before that, we need to understand what drives (or prevents) users from making progress toward these successful outcomes. In the next chapter, we'll explore the four forces that influence whether users will push through the onboarding journey or stick with their current approach.

8

The Four Forces of Progress

~~~~~

**People don't buy products; they hire them
to make progress in their lives.**

Bob Moesta

~~~~~

Imagine you're standing at the foot of a bridge. Behind you lies the familiar shore—your current solution, with all its known frustrations and limitations. Ahead, across the span, is the promise of something better—a new product that could transform how you work. What makes you decide to take that first step onto the bridge?

All these thoughts start racing through your mind:

- Will the new product really be better than my current solution?

- How much time will it take to learn?

- What if my team resists the change?

- What if we lose data in the transition?

These doubts are natural—they're part of every decision to change. To successfully onboard users, you need to help them see the value in making the journey across that bridge.

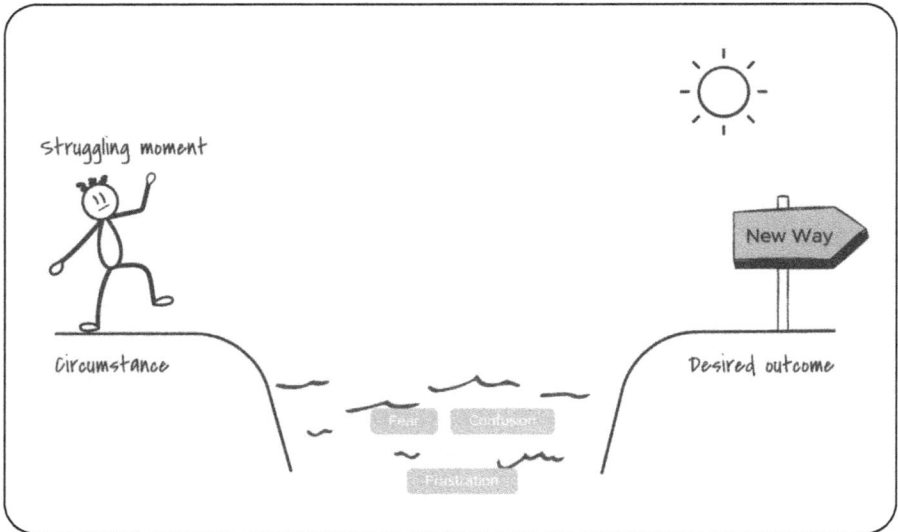

In this chapter, we'll explore the Four Forces of Progress framework developed by Bob Moesta and Chris Spiek at The Re-Wired Group to understand switching behavior.

To effectively apply this framework, we first need to understand what influences a user's decision to change solutions. Let's examine the four forces that drive users toward—or hold them back from—adopting your product.

THE FOUR FORCES OF PROGRESS

Once we understand what users might switch from, we need to examine what influences their decision to change. That's where the Four Forces of Progress come in.

Two forces drive users toward change:

- The **push** of frustrations with their current solution
- The **pull** of potential benefits from your solution

Two forces hold users back:

- The **inertia** of existing habits and workflows
- The **anxiety** about the risks and uncertainties of changing

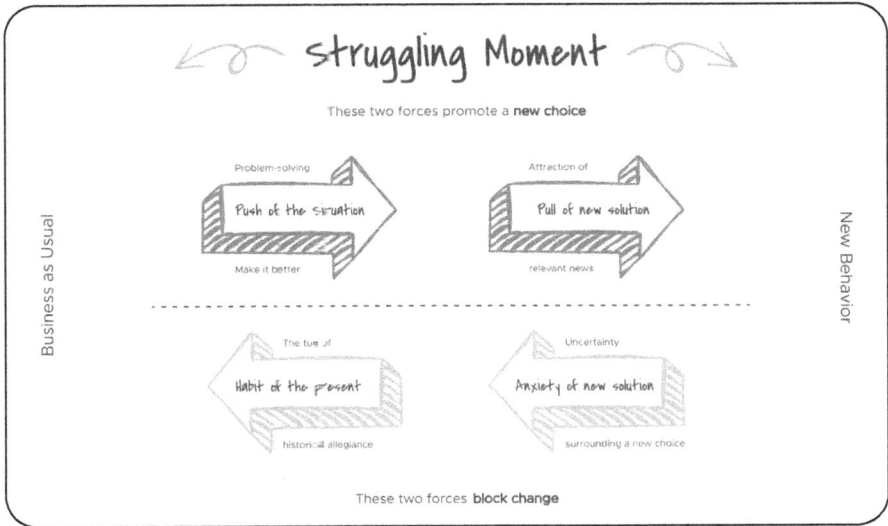

Understanding and managing these forces is crucial for successful onboarding. It's not enough to build great features or create helpful guides—you need to actively address what's driving users toward (and holding them back from) adopting your solution. Let's examine each force in detail.

Push: The Problem with the Current Solution

The push force represents the frustrations and problems that drive users away from their current solutions. These pain points create mounting pressure until users finally decide they need to make a change.

Common push forces might include:

- Slow or unreliable performance
- Missing essential features
- Poor user experience
- High costs or unexpected fees
- Limited scalability
- Integration difficulties
- Time-consuming manual work
- Lost or disorganized information

While teams might tolerate these frustrations for months, it often takes a critical moment—like losing a major client or hitting system limitations—to finally drive action. And even then, push forces alone won't drive change; they must be strong enough to overcome both anxiety about switching and the comfort of familiar routines. That's why successful products help users clearly see the cost of sticking with their current solution.

For your product, ask yourself:

- What frustrates users most about their current solution?
- What limitations do users frequently hit with their current tools?
- What costs (time, money, opportunities) do they incur by maintaining the status quo?

By articulating these frustrations clearly in your onboarding, you can help users recognize why staying with their current solution is more painful than the effort of switching to yours.

Pull: The Attraction to the New Solution

The pull force represents your product's magnetic appeal. It draws users toward a better future—the positive outcomes and improvements they hope to achieve by adopting your product.

Let's return to our team using spreadsheets for project management. When they discover a dedicated project management tool, they're pulled by possibilities like:

- Real-time collaboration without version control issues

- Automated workflows for routine tasks

- Visual dashboards for instant progress tracking

- Integrated communication for contextual discussions

These benefits create a compelling vision of the future. Teams imagine spending less time in status meetings and more time on getting things done. They see themselves delivering projects consistently on time and on budget. They envision better visibility and control over their work for all project stakeholders.

As these pull forces strengthen, teams become more willing to overcome the effort of changing. But the pull must be stronger than both the anxiety about change and the comfort of current habits. That's why successful products paint a clear, achievable picture of life after adoption.

For your product, ask yourself:

- What specific improvements do users envision with your solution?

- What capabilities will they gain that they don't have today?

- What positive outcomes can they achieve more easily?

By weaving these pull forces throughout your onboarding, you create excitement and motivation that propel users forward. Each milestone becomes an opportunity to reinforce the transformative future ahead, helping users overcome initial setup friction and embrace the change your product promises.

Inertia: Habits of the Present

The **inertia force** represents the comfortable routines and familiar workflows that keep users anchored to their current solution. These aren't just preferences—they're deeply ingrained behaviors that have become second nature over months or years of use.

Even with all the spreadsheet frustrations we discussed earlier, our project management team remains anchored to their solution through several powerful habits:

- Muscle memory in using familiar shortcuts

- Established workflows that everyone understands

- Customized systems built over time

- Historical data embedded in existing files

These habits create a powerful resistance to change. Teams worry about losing their efficiency during the transition. They've invested time in creating templates and training people on their current process. What seems frustrating has at least become predictable and manageable.

As these inertia forces strengthen, teams become more hesitant to change, even when they see better alternatives. That's why successful products need to acknowledge and address these existing habits rather than dismiss them. The familiar, even with its flaws, often feels safer than the unknown.

For your product, ask yourself:

- What existing workflows and habits will users need to change?
- What investments in their current solution make them reluctant to switch?
- What familiar routines provide comfort despite their limitations?

Understanding these inertia forces will help you create an onboarding experience that respects existing workflows while gradually introducing new, better ways of working. By acknowledging what users are giving up, you can better guide them toward what they'll gain.

Anxiety: Uncertainties of Change

Unlike the other forces, the **anxiety force** is about the unknown—the fears, doubts, and uncertainties that surface when contemplating change. It's the voice in your head asking, "What if?" and imagining all the ways things could go wrong.

Our project management team's anxieties about switching to a new solution could show up in various ways:

- Data migration fears: "What if we lose historical project data in the transfer?"
- Team resistance: "How will we get everyone trained and aligned on the new system?"
- Integration concerns: "Will it work with our other critical tools?"
- Performance uncertainty: "Can we trust it to handle our complex workflows?"

These anxieties often lurk beneath surface-level objections. A team might say, "It's too expensive," when they're really thinking, "What if we invest all this time and money, and it doesn't work out?" They might claim, "We're too busy to switch now," when they're actually wondering, "What if the transition disrupts our current projects?"

The more significant the change, the stronger these anxiety forces become. That's why successful products don't just showcase features—they actively build confidence through social proof, phased implementations, and small wins early in the onboarding journey. By letting teams roll out changes gradually or test with a small group first, you reduce the perceived risk of the transition.

For your product, consider these anxiety-related questions:

- What specific risks and uncertainties worry your users most?
- Which past experiences make them skeptical of change?
- What assurances do they need to feel confident moving forward?

Your onboarding must systematically address these anxieties while building confidence through proven success patterns. When users see others like them succeeding, their path forward becomes less daunting.

UNDERSTANDING WHAT USERS SWITCH FROM

Now that we understand the forces at play, let's examine what users might be switching from. This knowledge is crucial because it helps us:

- Identify specific push forces that drive users away from current solutions

- Highlight distinctive pull forces that make your solution attractive

- Anticipate the inertia of existing workflows users need to break

- Address anxieties specific to each type of switch

Most companies focus only on direct competitors, but users aren't just choosing between similar solutions—they're choosing between different ways of getting their job done. These alternatives fall into three categories, from closest to furthest from your solution:

1. **Direct Competitors:** Products that solve the job the same way as yours.

2. **Secondary Competitors:** Different approaches to solving the same job.

3. **Indirect Competitors:** Solutions that eliminate the need for the job.

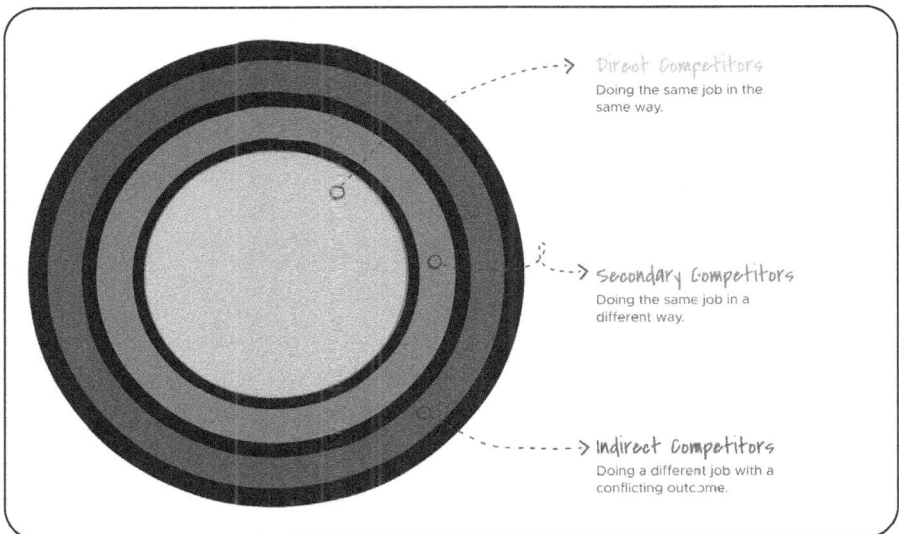

Direct Competitors
Doing the same job in the same way.

Secondary Competitors
Doing the same job in a different way.

Indirect Competitors
Doing a different job with a conflicting outcome.

Let's look at how this applies to a project management tool:

Direct Competitors: Other project management platforms like Asana, Monday.com, or ClickUp that offer similar features and approaches to organizing work.

Secondary Competitors: Different ways teams manage their work: spreadsheets and shared documents, email threads and chat channels, physical whiteboards and sticky notes, and custom-built internal tools.

Indirect Competitors: Solutions that eliminate the need for project management software: hiring dedicated project coordinators, restructuring into self-managing teams, and outsourcing project work to agencies.

Understanding these alternatives reveal:

- The specific pain points driving users away from each type of solution

- The unique value your product offers compared to each alternative

- The established habits and workflows users will need to change

- The transition concerns you'll need to address

By mapping these alternatives against the four forces, you can create an onboarding experience that directly addresses the real challenges and concerns users face when switching from each type of solution.

BALANCING THE FORCES FOR SUCCESS

The four forces aren't just abstract concepts—they're practical tools for driving real adoption. When these forces are properly balanced, users naturally progress toward your solution. When they're not, even the best product features can fail to gain traction.

In the next chapter, we'll build on this understanding with the User Success Canvas Exercise. Working with your onboarding team, you'll map out the three dimensions of user success—functional, emotional, and social—from the previous chapter and examine how these four forces influence progress toward that success. This exercise will help you create an onboarding experience that not only acknowledges these forces but actively manages them throughout the user's journey.

EUREKA ACTION ITEMS: THE FOUR FORCES OF PROGRESS

Take some time to analyze the four forces affecting your product's onboarding. You can find a fillable template at *eurekabonus.com*, but start by answering these questions:

Step 1: Map Your Competition

List out what users might be switching from:

- **Direct Competitors:** Similar products solving the same job

- **Secondary Competitors:** Different approaches to the same job

- **Indirect Competitors:** Solutions that eliminate the need

Step 2: Identify Forces Driving Progress

- **Push Forces:** What frustrates users about their current solution?

- **Pull Forces:** What attracts users to your solution?

Step 3: Recognize Forces Blocking Progress

- **Inertia Forces:** What habits and workflows keep users in place?

- **Anxiety Forces:** What risks and uncertainties worry users?

Use these insights to shape your onboarding strategy. Remember: Successful adoption requires both strengthening the forces that drive progress (push and pull) and reducing those that block it (anxiety and inertia).

9

The User Success Canvas Exercise

～～～

**Success is not about the destination,
it's about the transformation.**

John Maxwell

～～～

In Chapters 7 and 8, we explored the three components of user success and the four forces that influence progress. Now, it's time to combine these concepts in a practical exercise that will help your team align on what success looks like for your users and what drives them toward (or away from) that success.

THE USER SUCCESS CANVAS EXERCISE

The User Success Canvas is a collaborative workshop that helps teams identify and document:

- Key Jobs-to-Be-Done (JTBD) statements
- The three components of user success for each job
- The forces that enable or block progress

Download a digital copy of the User Success Canvas at *eurekabonus.com*.

OBJECTIVE

The User Success Canvas Exercise helps teams identify and align on:

- What success truly means for their users across functional, emotional, and social dimensions
- Which forces drive users toward (or away from) that success
- How different user segments and roles define success differently
- Where to focus onboarding improvements for maximum impact

By the end of this exercise, your team will have a shared understanding of user success and a prioritized list of opportunities to improve your onboarding experience.

A NOTE ABOUT USER RESEARCH

This exercise assumes you have Sales and Customer Success teams that regularly interact with prospects and customers. These team members bring valuable insights about common objections, challenges, and success stories they hear daily.

To get the most value from this exercise, have your team:

- Review 3–5 recent sales call recordings and 3–5 customer onboarding calls.
- Collect common objections and challenges from prospects.
- Gather success stories from recently onboarded customers.

If your team doesn't have regular customer interactions (e.g., your company is purely a self-serve product), conduct user interviews and customer research before running this workshop. Review the findings with your team before the exercise. You can find interview guides and user research templates at *eurekabonus.com*.

THE SETUP

The exercise uses a collaborative canvas format to help teams visualize all aspects of user success—from Jobs-to-be-Done statements to forces affecting progress. This visual approach makes it easy for team members to contribute their perspectives and create a shared understanding of your onboarding challenges.

Draw or display the User Success Canvas, dividing it into these key sections:

- Jobs-to-be-Done statements.

- The three dimensions of success: Draw a human head facing right. Place Functional Success (see) on the right side where the eyes are, Emotional Success (feel) at the bottom where the heart would be, and Social Success (hear) on the left near the ears. This visual representation helps teams understand how users experience success through different senses.

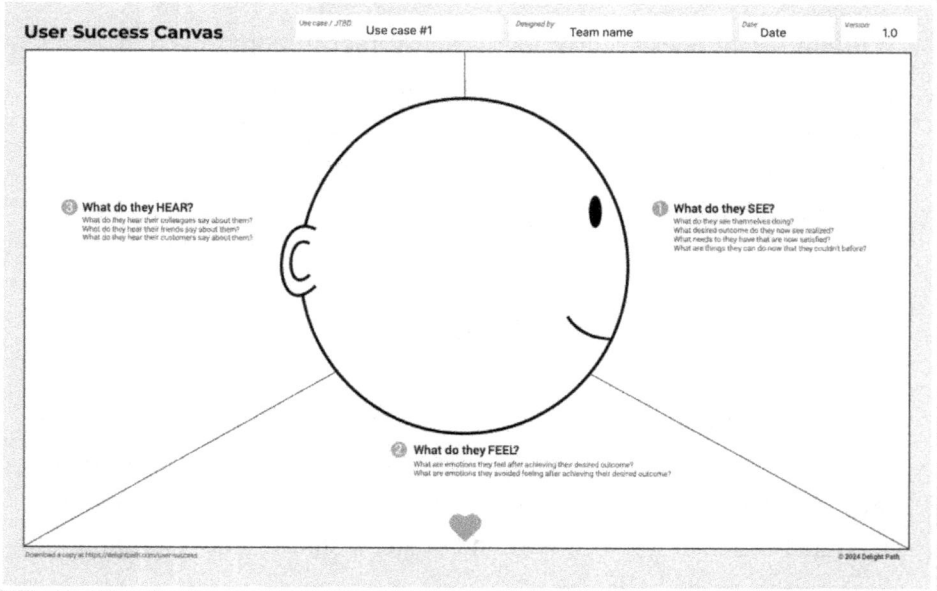

- Forces that enable or block progress: Create four areas—two on top (Push and Pull forces that drive progress) and two on the bottom (Inertia and Anxiety forces that block progress). This layout helps teams visualize how different forces either propel users forward or hold them back.

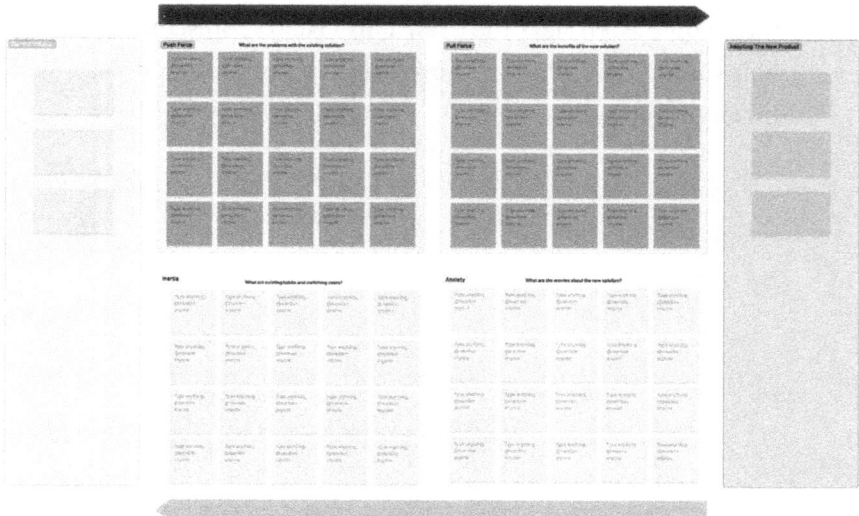

Time needed: 45–60 minutes

Participants: Ideally, this exercise should include the same cross-functional team (four to eight people) who participated in the Sailboat Exercise from Chapter 6. Having the same team members ensures continuity and builds on the shared insights already developed about your onboarding challenges.

Materials needed:

- Whiteboard or digital whiteboard (FigJam and Miro templates available at *eurekabonus.com*)

- Square sticky notes (different colors for different sections)

- Sharpies or markers

- Voting dots (3 per participant)

- Timer

- Your completed User Success Canvas(es)

You can also use a template I've created for FigJam and Miro. You can access it at *eurekabonus.com.*

FACILITATION TIPS

Whether running this exercise in person or remotely, keep these key tips in mind:

- Use different-colored sticky notes (or digital equivalents) for each section.

- Keep timeboxes strict and use visible timers.

- Encourage equal participation through silent writing phases.

- For remote sessions, use digital tools like Miro or FigJam.

- Document the output through photos or screenshots.

- Have all materials and tools prepared in advance.

Download detailed facilitation guides for both in-person and remote sessions at *eurekabonus.com*.

THE EXERCISE

Part 1: Jobs-to-be-Done Statements (15 minutes)

As we explored in Chapter 7, B2B products typically serve multiple stakeholders—from executives concerned with ROI to end users focused on daily tasks. The first step is identifying the key Jobs-to-be-Done (JTBD) statements that describe why different users hire your product.

For each key stakeholder role (buyer, end-user, admin, etc.), you'll want to capture their jobs in a structured format that shows:

- The situation that triggers their need (When . . .)
- Their motivation driving the change (I want to . . .)
- Their expected outcome (So I can . . .)

For example, a JTBD statement for a project management tool might be: "When managing multiple team projects, I want to keep everything organized and visible so I can deliver results on time and keep stakeholders informed."

Different roles might have very different JTBD statements. While a team member might focus on daily task management, an executive might care more about resource allocation and reporting.

Let's begin:

1. Give each participant sticky notes and a marker.
2. Set a timer for 3–5 minutes. Ask participants to silently write down JTBD statements they've heard from customers using the format:
    o "When I [situation], I want to [motivation], so I can [expected outcome]."
    o Encourage them to write statements for different roles they interact with.
    o Example for buyer: "When evaluating team productivity tools, I want to see clear ROI metrics so I can justify the investment to finance."
    o Example for end-user: "When managing my daily tasks, I want quick access to priorities so I can focus on what matters most."
3. Have everyone stick their notes on the board randomly.
4. As a group, organize similar statements together.
5. Give each participant 3 voting dots.
6. Set a timer for 3 minutes and explain the voting rules:
    o They can vote on their own sticky notes.
    o They can put multiple votes on one sticky.
    o They must use all their votes.
    o No talking during voting.
    o Vote for the JTBD statements that best represent your product's core value and most common user needs based on customer interactions.

7. Once time is up, have everyone sit down.

8. Create a vertical stack of sticky notes, ordered from most votes to least.

9. The top 2–3 voted JTBD statements represent your primary user journeys and will be your focus for mapping success criteria in the rest of this exercise.

Part 2: The Three Dimensions of User Success (20 minutes)

Now that you have your key JTBD statements, you'll guide your team through exploring user success across three dimensions. Remember that success isn't just about functional outcomes—it includes emotional satisfaction and social validation.

You'll help your team understand:

- Functional Success (See): What new capabilities your users gain

- Emotional Success (Feel): How your users feel after achieving success

- Social Success (Hear): What others say about your users

You'll spend about 5 minutes on each dimension, helping your team build a complete picture of what success looks like for your users.

For each JTBD statement, spend 3–5 minutes on each component:

1. Functional Success (5 minutes)

- Give participants yellow sticky notes and markers.
- Set a timer for 3 minutes of silent writing.
- Ask them to write what users can now do that they couldn't before.
- One idea per sticky note using the format: "I can now [action/capability]"
- Have everyone post their notes in the "see" area of the head.

Take a few minutes to review as a group:

- What patterns do you notice in the responses?
- Does anything surprise you?
- Are there capabilities or outcomes we might be missing?
- Which items resonate most strongly with the group?

2. Emotional Success (5 minutes)

- Give participants pink sticky notes.
- Set a timer for 3 minutes. Ask them to write:
- Positive emotions users feel after success
- Negative emotions they avoid
- One idea per sticky note using the format: "I now feel [positive emotion]." and "I now don't feel [negative emotion]."
- Place notes in the "feel" area at the bottom of head.

Pause to discuss:

- Which emotions came up most frequently?
- Are there any unexpected emotional impacts?
- Does this align with what we hear from customers?

3. Social Success

- Give participants green sticky notes.
- Set a timer for 3 minutes. Have them write statements in the format:
  - "[Stakeholder]: [What they say about the user]."
  - Place notes in the "hear" area near the ears.

Review together:

- Which stakeholders appear most frequently?
- What themes emerge in how others perceive success?
- Are we missing any important perspectives?

Part 3: Forces of Progress (20 minutes)

Now that you've mapped user success, you'll help your team examine what drives users toward (or away from) that success. Guide them in exploring the four forces that influence progress:

- Push: What frustrates your users about their current solution?
- Pull: What attracts your users to your solution?
- Inertia: What familiar routines keep your users in place?
- Anxiety: What risks worry your users about changing?

You'll lead two focused sessions: first, exploring what enables progress (push and pull forces), then examining what blocks it (inertia and anxiety). This will help your team understand both what motivates users forward and what holds them back.

You can download FigJam and Miro templates for the Four Forces of Progress exercise at *eurekabonus.com*.

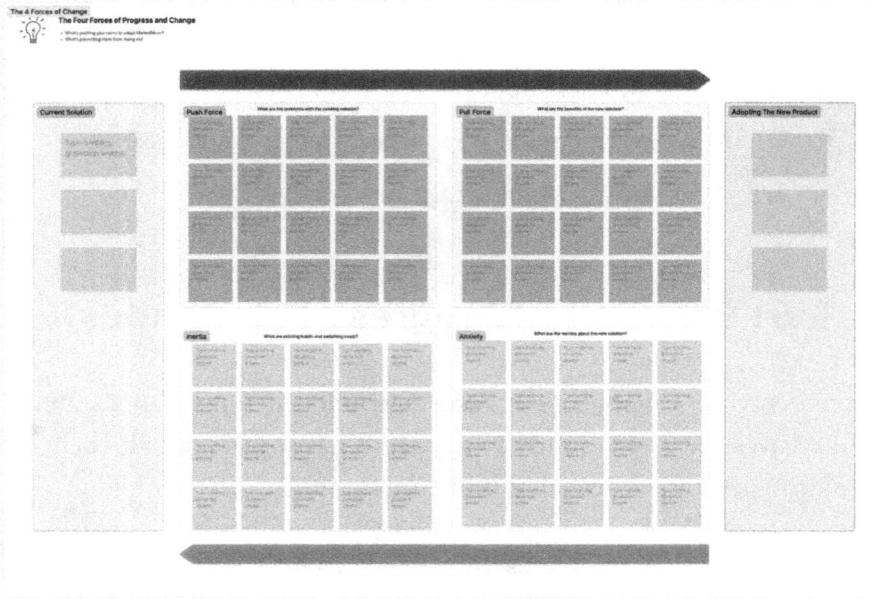

Session 1: Forces that Enable Progress (10 minutes)

- Give your team green sticky notes.
- Set a timer for 5 minutes of silent writing.
- Have participants write one idea per sticky note about:
  - Push: What frustrates users about their current solution?
  - Pull: What attracts users to your solution?
- Have them place Push forces in the top left quadrant and Pull forces in the top right.
- Spend 5 minutes discussing as a group:
  - What patterns do you notice?
  - Does everyone agree with the forces identified?
  - Are there different perspectives to consider?

Session 2: Forces that Block Progress (10 minutes)

- Give your team pink sticky notes.
- Set a timer for 5 minutes of silent writing.
- Ask them to write one idea per sticky note about:
  - Inertia: What existing workflows or investments make change difficult?
  - Anxiety: What risks or uncertainties worry users?
- Place notes in the corresponding areas: inertia forces on the bottom left and anxiety forces on the bottom right.
- Spend 5 minutes discussing as a group:
  - What surprises emerged?
  - Are there forces we might be underestimating?
  - How do different roles perceive these blocking forces?

Part 4: Repeat for Additional Jobs-to-be-Done

For each additional top-voted JTBD statement from Part 1:

1. Take a short break (5–10 minutes).
2. Create a new canvas.
3. Repeat Parts 2 and 3, focusing on this specific role or use case.

Remember to consider what success looks like for different stakeholders:

- Buyers vs. end users (e.g., IT decision-makers vs. daily users)
- Departmental roles (Marketing, Sales, IT)
- Seniority levels (Individual contributors vs. Managers)

Look for patterns across these different roles—they often reveal opportunities for creating shared value in your onboarding experience.

OUTPUT AND NEXT STEPS

The completed User Success Canvas provides:

- Clear JTBD statements to guide onboarding
- Success criteria across functional, emotional, and social dimensions
- Understanding of forces influencing user progress
- Research priorities based on key assumptions

In the next steps, we'll use these insights to:

- Design targeted onboarding experiences
- Create success metrics that matter to users
- Address key anxieties early in the user journey
- Build confidence through social proof
- Maintain momentum by reducing friction

EUREKA ACTION ITEMS: THE USER SUCCESS AND FORCES OF PROGRESS WORKSHOP

Ready to run your own User Success Canvas workshop? Download the template and facilitation guide from *eurekabonus.com*. Remember, this is an iterative process—your understanding of user success will evolve as you gather more customer insights.

In the next chapter, we'll reverse-engineer this journey, turning these insights into actionable onboarding experiences that guide users toward their desired outcomes. We'll explore how to create momentum by addressing blocking forces early while amplifying the forces that drive progress forward.

STEP 3

Reverse Journey Map to Success

Establish a Team

Understand Success

Reverse Journey Map

Keep Users Engaged

Apply Analyze and Repeat

10

The Ultimate Win for the Customer

~~~

**Begin with the end in mind. He who has a
"why" to live can bear almost any "how."**

Friedrich Nietzsche

~~~

For my wife's thirtieth birthday, I planned a special road trip from
Toronto to Boston. She loves lobster, so the destination was an
easy choice. Before we left, I plugged our hotel's address into Google
Maps and set off on what should have been a straightforward journey.
Somewhere along the way, I missed a crucial turn. Because of the
distance between exits, what should have been a quick correction
turned into a thirty-minute detour.

But here's the thing—Google Maps immediately recalculated our
route. It didn't just tell us where I went wrong; it created a new path
to our destination. We still reached our hotel, just a bit later than
planned, and Joanna still got her birthday lobster dinner.

In the same way, we need both a clear destination and key milestones
to guide our users. In the previous step of the EUREKA Framework,
we explored the destination of user success. Now, it's time to identify
those crucial waypoints—the pillars of the bridge that will support

users on their journey from their current circumstances to their desired outcome.

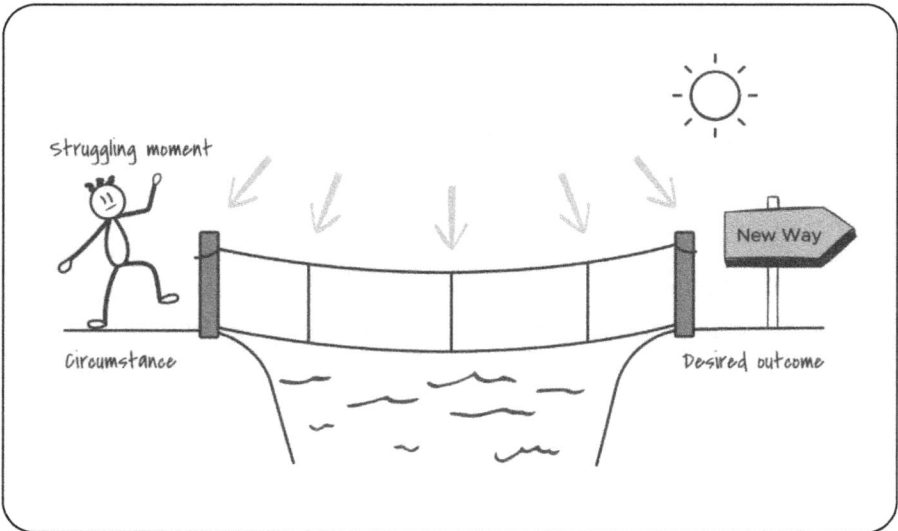

Like a well-planned journey, we'll start with the destination and work backward to create a clear path to success. In this chapter, we'll explore how to define your customer's Ultimate Win—that crucial first milestone where customers experience meaningful value from your product. Getting customers to start their journey is only half the battle—you need to guide them to a destination worth reaching.

WHAT IS THE ULTIMATE WIN?

The Ultimate Win is an early indicator of user success, which we discussed in Chapter 7. While some call it the "aha moment" or "activation moment," I purposely chose the term Ultimate Win because it represents something deeper than feature adoption or completing setup tasks. It's the moment users cross the onboarding bridge and catch their first glimpse of the transformation ahead—truly validating their decision to change.

Five characteristics distinguish a product's Ultimate Win:

1. The Ultimate Win is a victory for the user

Many teams track "selfish metrics" like "completes profile setup" or "configures integration"—actions that benefit the product but not the user. I saw this while consulting with a construction photo tool. They set "takes ten photos" as their activation moment because data showed these users were more likely to stay active. But the real Ultimate Win wasn't taking photos—it was getting client approval through those photos, eliminating the need for site visits.

The takeaway? **A true Ultimate Win must represent a genuine victory for your users.** It's not just a product usage milestone. When you get this right, it naturally becomes a win-win for both the customer and your company.

To find your product's true Ultimate Win, use the "Five Whys" exercise—keep asking "why" until you uncover the deeper motivation behind each action. For example:

- Why should contractors give access to their mobile phone cameras? To take pictures.

- Why do they need to take pictures? To share updates and issues with clients.

- Why do they need to share updates? To get feedback and approval.

- Why do they need feedback and approval? To protect themselves from liability.

- Why do they need liability protection? To run their business confidently and securely.

See how the Ultimate Win emerges? It's not about taking photos—it's about getting verification that protects their business. This leads us to our second characteristic.

2. The Ultimate Win is an early indicator of success

While the full transformation we discussed in Chapter 7 takes time, the Ultimate Win serves as an early signal that users are on the right path. Like the first successful meeting with a project management tool or the first scheduled appointment through a calendar app, these moments validate that the user's decision to change is paying off. It's often predictive of long-term retention because users have experienced, not just understood, your product's core value.

3. The Ultimate Win needs to be trackable

Your Ultimate Win must be something you can measure and track consistently. It becomes a key milestone in your onboarding experience and a critical metric for success. When someone asks how successful your onboarding experience is, you should be able to say, "Eighty-eight percent of our users achieve the Ultimate Win within their first week." This measurability helps you identify where users get stuck and optimize your onboarding accordingly.

4. The Ultimate Win should be repeatable

The Ultimate Win isn't a one-time achievement—it's an experience users should want to repeat frequently. Each repetition strengthens their product habits and deepens their transformation. For example, when a project manager runs their first status meeting using real-time data from their project management tool, and everyone leaves aligned, they'll want to repeat this success with other meetings and projects. The more they achieve this Ultimate Win, the more they build a habit of data-driven project management and transparent communication. This repeatability transforms occasional victories into sustained behavior change.

5. The Ultimate Win should be simple

Facebook's famous "aha moment"—adding seven friends in ten days—didn't come easily. Their data science team analyzed billions of user interactions, testing countless combinations of actions and timeframes to find this precise correlation with long-term retention.

While this data-driven approach works for companies with millions of users and sophisticated data science teams, most B2B companies don't have enough data or the expertise to find such precise correlations. That's why I advise the companies I work with to focus on something simpler: Identify your Ultimate Win (which we'll talk about in Chapter 12) and remove any roadblocks preventing users from achieving it.

A focused approach works better than optimizing for multiple metrics or complex combinations of actions. Understanding what represents a meaningful win for your users allows you to direct all onboarding efforts toward that one crucial milestone.

EXAMPLES OF ULTIMATE WINS

Let's look at how different B2B products define their Ultimate Win by connecting it to their users' deeper transformation goals:

For a project management tool like Asana:

- User Success: Transforming from an overwhelmed coordinator into a confident project orchestrator who predictably delivers successful outcomes
- Ultimate Win: When another teammate completes their first task in a shared project, validating collaborative project management

For a calendar scheduling tool like Calendly:

- User Success: Becoming known as a highly efficient professional who respects everyone's time
- Ultimate Win: First meeting booked through their scheduling link

For a sales CRM like Salesforce:

- User Success: Evolving into a data-driven sales leader who consistently exceeds targets
- Ultimate Win: First accurate pipeline forecast using CRM data

For a design tool like Canva:

- User Success: Becoming the go-to person for creating professional-looking content without needing design expertise
- Ultimate Win: First design shared and praised by the team

DEFINING YOUR ULTIMATE WIN

Several approaches can help you identify your product's Ultimate Win. Each provides a different perspective: Customer interviews reveal qualitative insights, usage data shows quantitative patterns, and team workshops leverage your organization's collective wisdom.

1. Learn from Customer Conversations

Talk to your power users and conduct Jobs-to-be-Done interviews to uncover their journey. Ask them about key moments: When did they first feel confident they made the right choice? You'll find interview guides and templates at *eurekabonus.com*.

Customer onboarding calls provide another valuable opportunity to identify your Ultimate Win. During these calls, watch for what Wes Kao, the co-founder of Maven.com, calls the **"Eyes Light Up" (ELU) moment**—that instant when customers become visibly excited while using your product. You might notice them perking up, leaning forward, or interrupting with enthusiastic comments.

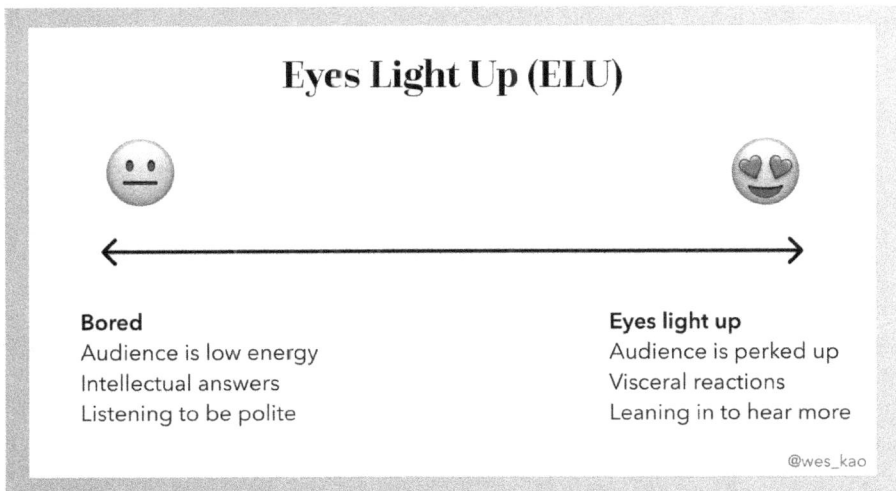

Eyes Light Up (ELU)

Bored
Audience is low energy
Intellectual answers
Listening to be polite

Eyes light up
Audience is perked up
Visceral reactions
Leaning in to hear more

@wes_kao

These ELU moments during customer onboarding calls are possible indicators of your Ultimate Win. When you notice patterns in what triggers these ELU moments across multiple calls, you've likely identified a key milestone in your user's journey.

2. Analyze Usage Patterns

Product analytics tools can help identify patterns among your most successful users. For example, Amplitude's Compass Chart feature specifically helps discover potential Ultimate Wins by analyzing the correlation between early user actions and long-term retention. The tool generates a heat map showing which actions and when they occur are most predictive of seven-, thirty-, sixty-, and ninety-day user retention.

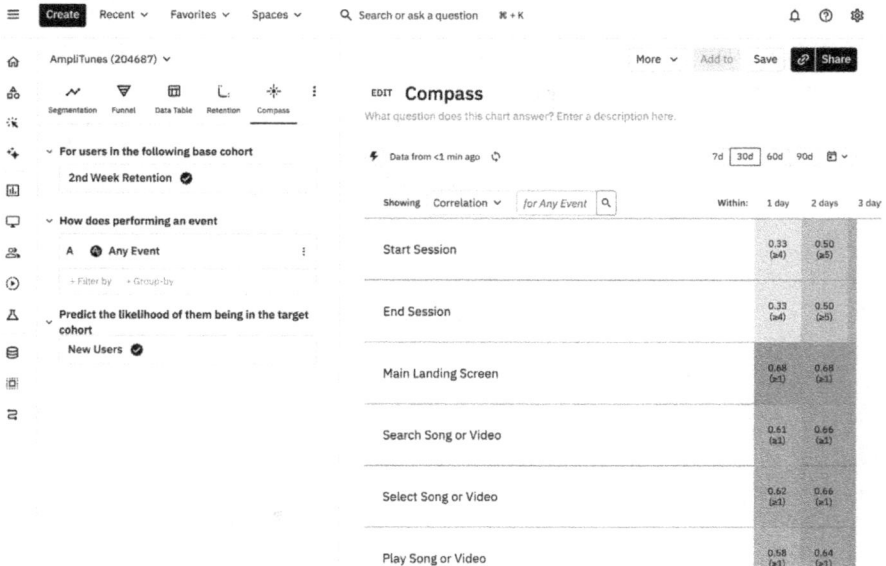

While tools like Pendo, Mixpanel, and Heap offer similar correlation analysis features, keep in mind that most B2B products don't often acquire enough new users to generate statistically significant correlations. If you're seeing fewer than one hundred new users per month, focus on qualitative insights from customer interviews and onboarding calls rather than trying to find statistical patterns in your usage data.

3. Workshop with Your Team

The wisdom of the crowd can be particularly powerful when defining your Ultimate Win. Each team brings valuable perspectives:

- Sales knows which features trigger prospects to switch from competitors.
- Customer Success sees Eyes Light Up moments during onboarding calls.
- Product has retention data and feature usage patterns.
- Support understands common pain points and success stories.

Getting these teams together not only combines these diverse insights but also creates the cross-functional alignment we discussed in Chapters 4 and 5. When every team contributes to defining the Ultimate Win, they develop shared ownership of the outcome—transforming it from a product metric into a company-wide mission. This collaborative approach helps prevent the "broken telephone" effect we explored earlier, where success definitions get distorted as they pass between departments.

Don't let perfect be the enemy of good—defining your Ultimate Win doesn't have to be a lengthy process. It's a hypothesis you can refine over time as you learn more about your users. The key is to start somewhere and give your team a clear target to optimize toward.

In Chapter 12, we'll explore a structured workshop activity that helps you identify your Ultimate Win and create a Reverse Journey Map. This exercise will help you leverage your team's collective wisdom while maintaining focus on what truly matters to your users.

COMMON PITFALLS TO AVOID

When defining your Ultimate Win, watch out for these common mistakes:

1. Choosing "Selfish Metrics"

Many companies define their Ultimate Win based on what they want users to do rather than what represents a meaningful victory for users. They focus on metrics like "completes profile setup" or "invites team members" because these actions benefit the company's growth. But if these actions don't deliver immediate value to users, they're not true wins.

2. Setting the Bar Too Low

Don't settle for surface-level actions that don't represent meaningful value for users. The Ultimate Win should be an early indicator that users are on the path to experiencing the transformation they hired your product to enable. Use the "Five Whys" exercise to dig deeper and find that meaningful moment where users first glimpse their desired transformation.

3. Overcomplicating the Definition

While you shouldn't set the bar too low, keep your Ultimate Win simple. For a project management tool, instead of tracking complex combinations of tasks created, team members invited, and integrations set up, focus on one clear moment: When another teammate completes their first task in a shared project. Start with something clear and measurable—you can always refine it as you learn more about your users.

DIFFERENT JOBS, DIFFERENT ULTIMATE WINS

For B2B products, it's crucial to recognize that different user segments need different Ultimate Wins. If you identified multiple Jobs-to-be-Done in the previous step, each job needs its own unique Ultimate Win.

Take Wave Apps, a financial tool for entrepreneurs. Their team identified three distinct customer jobs as outlined in their signup process:

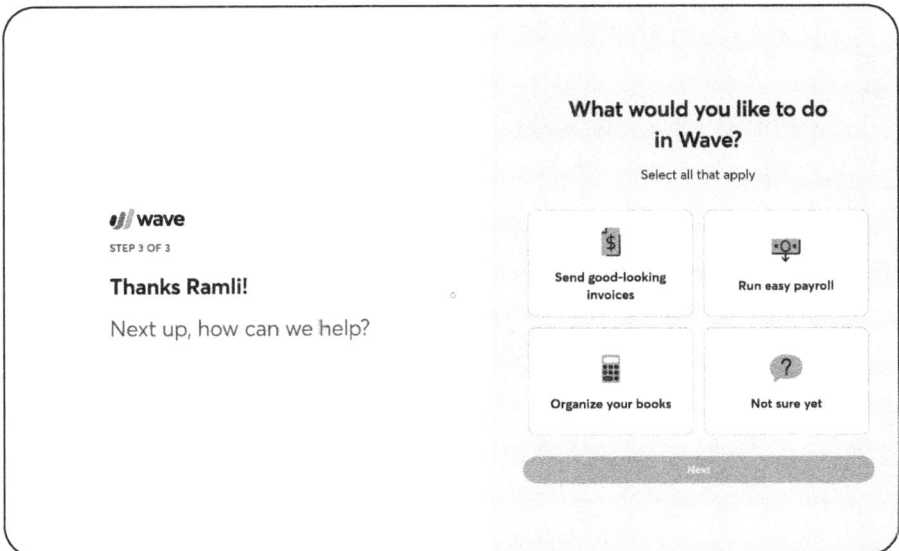

1. Send Professional Invoices

- User Success: Becoming known for running a professional, organized business
- Ultimate Win: When a user sends their first invoices using Wave

2. Manage Business Accounting

- User Success: Feeling confident and in control of business finances
- Ultimate Win: When they first reconcile their monthly transactions and understand their cash flow

3. Run Employee Payroll

- User Success: Being a reliable employer who pays people correctly and on time
- Ultimate Win: When they successfully complete their first payroll run with direct deposits

Each JTBD represents a different user need and transformation goal. This multi-win approach helps you:

- Create targeted onboarding paths for different user segments
- Set appropriate success metrics for each use case
- Deliver more relevant guidance and support
- Measure activation rates accurately by JTBD

To map your Ultimate Wins, group users by their primary JTBD and identify the earliest moment of meaningful value for each job. This ensures every user gets the most relevant path to success.

DESIGNING FOR THE ULTIMATE WIN

The Ultimate Win represents a crucial victory that validates their decision to change. Make sure it's:

- A genuine victory from the user's perspective
- An early indicator of deeper transformation
- Something trackable and measurable
- Repeatable and habit-forming
- Simple enough to achieve

Start with user success and work backward. Don't aim for perfection—create a hypothesis your team can rally around and refine over time. Whether you have one Ultimate Win or several for different user segments, the key is understanding what represents meaningful value for your users.

Since users might take days or weeks to achieve their Ultimate Win, the next chapter explores how to build momentum with "Same-Day Wins"—quick victories that keep users engaged along the way.

EUREKA ACTION ITEMS: DEFINING YOUR ULTIMATE WIN

Now that you understand what makes a strong Ultimate Win, it's time to start identifying yours. While we'll do a comprehensive exercise in Chapter 12, here are some steps you can take now:

1. Review your User Success Canvas and ask:

- What transformation are users seeking?
- What would validate their decision to change?
- When might they first experience meaningful value?

2. List potential Ultimate Win moments:

- What actions correlate with long-term success?
- When do users first share results with others?
- What achievement makes them feel confident?

3. Test your candidates against the five characteristics:

- Is it a genuine win for users?
- Does it indicate future success?
- Can you track it consistently?
- Will users want to repeat it?
- Is it simple enough to achieve?

11

Mapping the Onboarding Success Roadmap

~~~~

Success is not a single achievement but a series of small wins, each building upon the last.

Maya Angelou

~~~~

In 1969, psychologist Edwin Locke published a groundbreaking paper that transformed our understanding of human motivation. His research revealed that not only did employees perform better with specific, challenging goals (rather than vague directives like "do your best"), but they were 12 percent more productive when they could see clear evidence of their progress—even small wins—compared to those who only saw end-of-month results.

Locke's research explains why video game designers break long quests into smaller missions. By making progress visible, they keep users motivated toward distant goals.

Unlike consumer apps that deliver immediate personal value, it often takes days or weeks for organizations to fully realize a B2B product's value because of necessary setup work, stakeholder coordination, and organizational changes. To keep B2B users motivated toward their Ultimate Win, we need a series of meaningful victories along the way. In this chapter, we'll create a clear roadmap of wins and milestones that guide users from their current situation to their Ultimate Win. We'll start by examining the Same-Day Win—the crucial first victory that builds immediate confidence. Then, we'll explore the key milestones and wins along their journey.

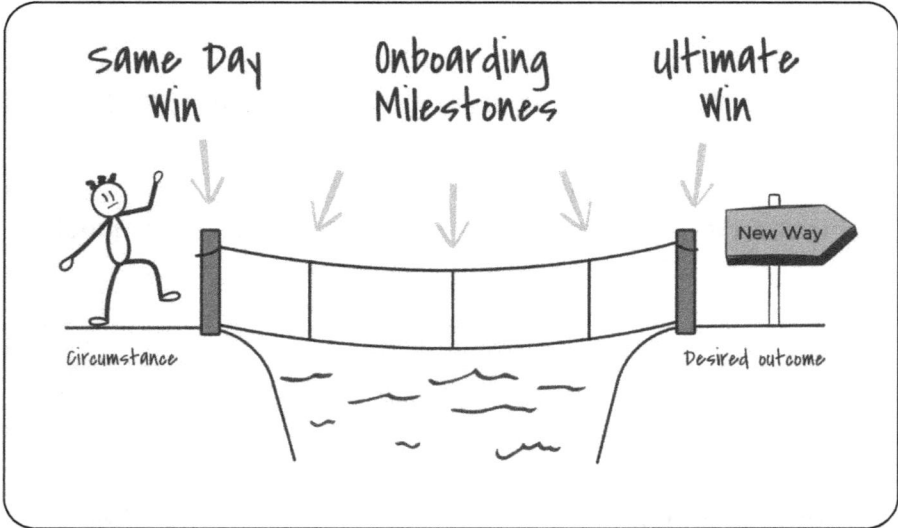

By mapping these strategic checkpoints, we'll create a path that keeps users engaged and confident throughout their transformation.

THE SAME-DAY WIN

The Same-Day Win is your user's first meaningful victory—ideally achieved within their first five to ten minutes of using your product. While others might call this a quick win, I emphasize "Same-Day" because it must happen during that crucial first session or moment of purchase when motivation and attention are highest.

The Importance of The Same-Day Win

The Same-Day Win is critical to B2B onboarding success for five key reasons:

1. Validates the Customer's Decision

B2B purchases involve multiple stakeholders and significant investment, which often triggers post-purchase anxiety (buyer's remorse). When champions who advocated for your solution achieve a meaningful win quickly, it validates their decision and provides concrete evidence they can share with stakeholders.

2. Reduces Emotional Friction

B2B implementations can feel overwhelming, with complex setup processes and organizational changes looming ahead. A well-designed Same-Day Win breaks through this anxiety by showing users they can achieve something valuable immediately—like getting a small taste of the destination before the journey begins.

3. Builds Momentum and Confidence

Remember Locke's research about visible progress? The Same-Day Win provides that crucial first victory that builds both product confidence and implementation momentum. Like taking the first step in a fitness journey, it creates positive energy that carries users through the more challenging tasks ahead. When users see they can be successful quickly, they're more likely to tackle the complex setup processes that follow.

4. Creates Internal Champions

Same-Day Wins produce immediate, shareable results that users can demonstrate to their teams. A customer success manager scheduling their first meeting through your calendar tool or a marketer previewing their first email campaign gets concrete wins they can showcase. This visible success helps turn early users into internal advocates who will champion your product throughout the longer implementation.

5. Accelerates Time-to-Value

While the Ultimate Win might be weeks away, Same-Day Wins deliver immediate business value. Even small victories—like importing your first customer list or creating a basic workflow—start generating returns on the investment immediately. This early value creation helps maintain organizational momentum and buy-in during the fuller implementation.

The Characteristics of a Same-Day Win

Every effective Same-Day Win shares three essential characteristics:

1. Achievable

The win must be easy to accomplish within minutes. For example, Calendly users can create their scheduling link in just three clicks—no complex setup required. The goal is to remove any friction that might prevent users from experiencing that first success.

2. Meaningful

The win should produce tangible results that matter to users, though this can take different forms:

- **Product Understanding**: Amplitude shows users sample reports with industry-specific dummy data, giving them immediate insight into the platform's value.

- **Progress Milestone**: For enterprise products, scheduling a kickoff call with a dedicated success manager can be meaningful, as it represents concrete progress toward implementation.

- **Quick Value**: Freshdesk users can experience the product by closing a sample support ticket and immediately understand how the system works.

3. Confidence-Building

The win should validate the user's decision and build trust in the product. Wave Apps offers a perfect example of this. When new users upload their company logo, Wave automatically extracts their brand colors and instantly shows a beautifully branded invoice.

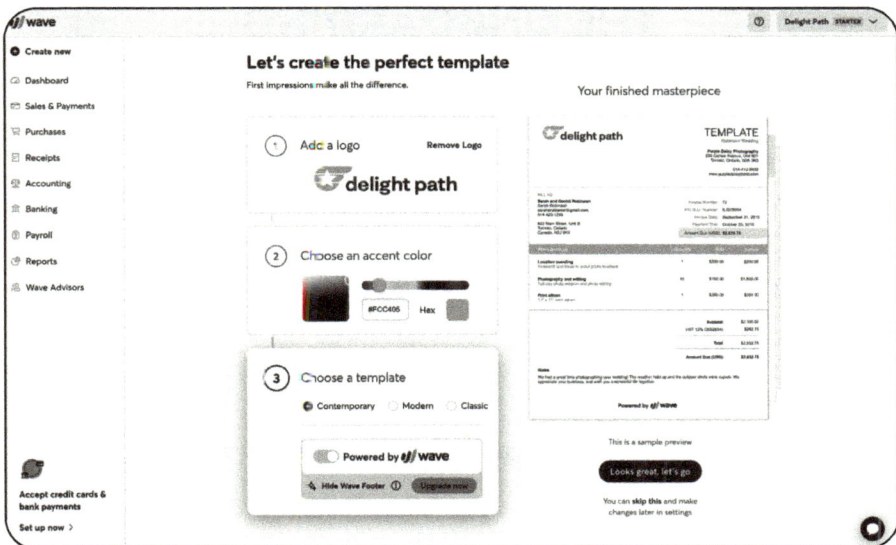

As Vivek Balasubramanian, Wave's former VP of Growth, observed:

"During customer interviews, the customers we talked to that saw what their invoice would look like with Wave said, 'Wow! This is great! This looks professional. It's beautiful.' That gives them a lot of confidence that the product is good. Wave is something that they can trust."

This confidence-building moment is crucial—it transforms the Same-Day Win from a simple task completion into a powerful affirmation of the user's purchase decision.

For your product, ask yourself:

- What's the smallest meaningful step toward your Ultimate Win?

- What early victory would validate their decision to choose your product?

THE ONBOARDING SUCCESS ROADMAP

With your Same-Day Win and Ultimate Win identified, you need clear milestones to guide users between these two endpoints. The milestones between them act like support pillars, confirming progress and maintaining momentum while preventing users from falling into confusion or doubt along the way.

Structuring Your Onboarding Wins

The most effective way to structure these wins is as follows:

[ACTION] the first [THING]

This format helps teams focus on specific, measurable achievements that mark clear progress. For example, here's how a project management tool might structure their onboarding journey:

Project Management Tool Onboarding Success Journey

1. Create the first project board (Same-Day Win).

2. Add the first team member.

3. Create the first task.

4. Assign the first task to a teammate.

5. Schedule implementation kickoff call.

6. Send the first task notification.

7. Review the first task update.

8. See teammate complete their first assigned task (Ultimate Win).

   Notice how each win builds on the previous one, gradually moving from basic setup tasks to more complex, value-generating activities. This progression serves several purposes:

9. Each win builds confidence through gradual progression.

10. Later wins often involve more team members, helping drive organizational adoption.

11. Users naturally discover more advanced features through progressive wins.

How to Identify Your Onboarding Journey Wins

Similar to the Ultimate Win discovery process in Chapter 10, you can identify your journey wins through three approaches:

1. JTBD Customer Interviews

Jobs-to-be-Done interviews with your power users can reveal the complete story of their journey. By understanding the timeline from their first trigger moment ("I need a better way to manage projects") through product selection and implementation, you'll discover the critical wins that kept them moving forward. These interviews help you map out:

- The key triggers that pushed them to seek a solution

- Their evaluation and decision-making process

- Which early product experiences validated their choice

- How they progressed from basic to advanced feature usage

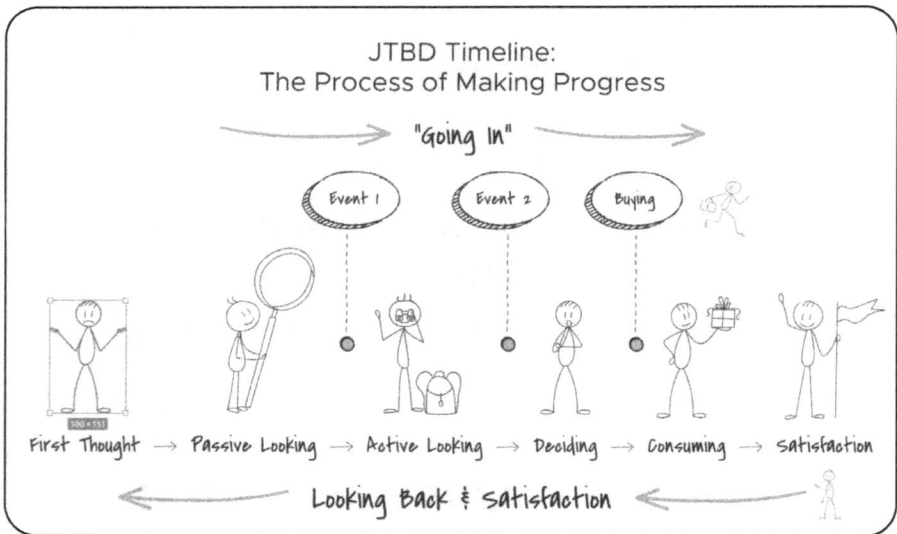

JTBD Timeline:
The Process of Making Progress

"Going In"

Event 1 — Event 2 — Buying

First Thought → Passive Looking → Active Looking → Deciding → Consuming → Satisfaction

Looking Back & Satisfaction

SOURCE: THE RE-WIRED GROUP

You'll find detailed guides and templates for conducting JTBD customer interviews at eurekabook.co/resources.

2. User Flow Analysis

Product analytics tools can reveal the common paths users take before reaching their Ultimate Win. These user flow reports show you which features new users engage with and in what sequence, helping you identify potential journey wins.

For example, Amplitude's Pathfinder report shows the sequence of actions (or "paths") users take during their early product experience. It can reveal the most common routes to conversion, helping you map out your onboarding milestones.

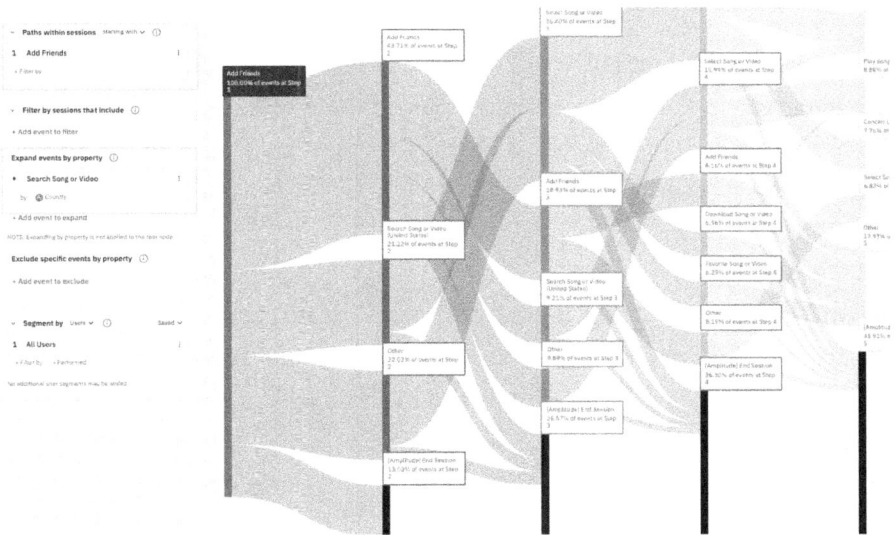

Mixpanel offers similar flow analysis tools that visualize user interaction sequences. These tools help you understand which features or actions typically precede successful activation.

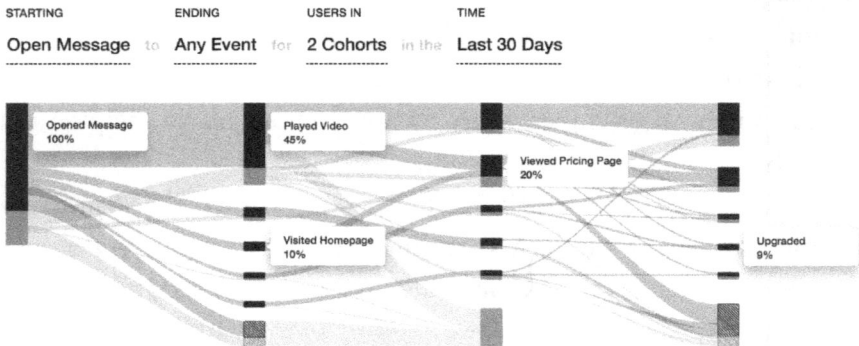

STARTING   ENDING   USERS IN   TIME

Open Message to Any Event for 2 Cohorts in the Last 30 Days

However, remember that B2B products often don't have enough users to generate statistically significant patterns. If you're seeing fewer than one hundred new users per month, prioritize insights from customer interviews over usage data.

3. Team Workshop

Your customer-facing teams witness the journey wins firsthand. Customer Success managers see multiple "Eyes Light Up" moments during onboarding calls—not just the Ultimate Win but all the smaller victories along the way. Sales know which quick wins to demonstrate in demos, and Support understands which early successes prevent users from getting stuck.

We'll explore how to bring these insights together with a reverse journey mapping exercise in the next chapter.

How Many Wins Should Your Onboarding Journey Have?

Most successful onboarding journeys include five to eight key wins between the Same-Day Win and Ultimate Win—clear milestones that mark meaningful progress, not small tasks. More than ten wins usually means you're tracking tasks rather than true milestones, which can overwhelm users and dilute the significance of each achievement.

The spacing between these wins is also crucial. Early wins should come quickly (hours or days apart), while later wins can be more spread out (weeks apart). This creates what game designers call "variable reward scheduling"—a powerful motivational technique that maintains engagement through varying intervals. For example:

- First two to three wins: Hours or days apart
- Middle wins: One to two weeks apart
- Final wins: Two to four weeks apart

Common Journey Mapping Pitfalls

When mapping out your onboarding journey, watch out for these common mistakes:

1. **Task-Based Instead of Value-Based**: Many teams focus on setup tasks that primarily benefit the company ("configure your profile" gets you user data, "invite teammates" helps with viral growth) rather than creating wins that deliver real value to users ("get your first qualified lead" or "close your first deal").

2. **Disconnected Journey**: Defining wins that don't clearly build toward the Ultimate Win. For example, if your project management tool's Ultimate Win is "when another teammate completes their first task in a shared project," having a milestone like "customize profile photo" or "set notification preferences" might feel productive but doesn't meaningfully progress users toward their goal.

Each milestone should feel like a logical step toward
the final destination—like "create first project board"
or "assign first task to teammate." Without this clear
connection, users can get lost in busy work that doesn't
move them toward success.

3. **Complex Early Wins**: Starting with wins that require too
    much organizational coordination or technical setup.
    For example, a CRM shouldn't make "import and clean
    entire customer database" their Same-Day Win because it
    takes too much time to complete and requires extensive
    data preparation, team coordination, and technical setup.

Instead, focus on something immediate and valuable like "create and
qualify first lead"—a simple win that still demonstrates the product's
core value. Complex early wins often lead to abandonment because
users lose momentum before experiencing any value.

DIFFERENT JOBS, DIFFERENT JOURNEYS

Just as we saw in Chapter 10 with Wave Apps's different Ultimate Wins,
each Job-to-be-Done requires its own unique onboarding journey. Let's
continue with the Wave Apps example to see how different jobs lead
to different Same-Day Wins and milestone paths:

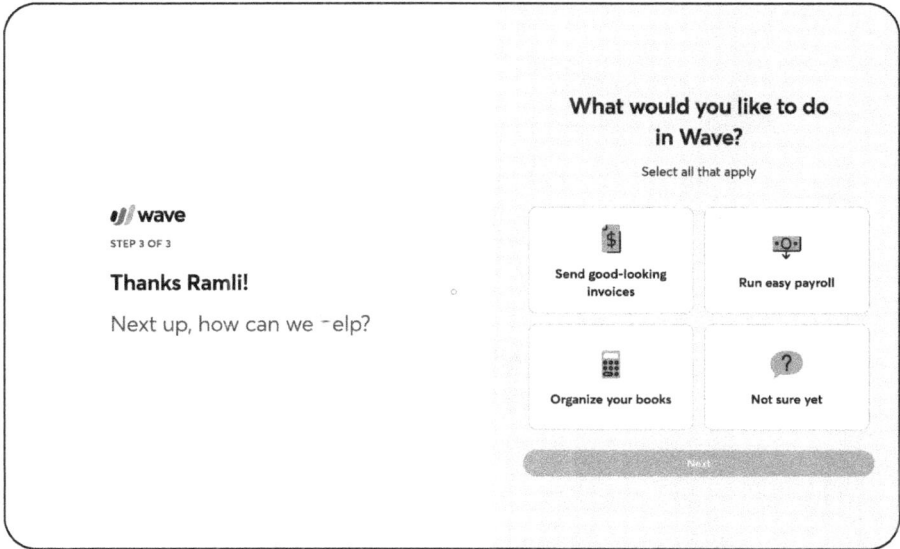

1. Invoicing Journey

- Same-Day Win: Preview first branded invoice

- Journey Milestones:
    - Import the first customer contact
    - Create the first invoice
    - Send their first invoice
    - Receive first online payment (Ultimate Win)

2. Accounting Journey

- Same-Day Win: Connect first bank account

- Journey Milestones:
    - Import the first transaction batch
    - Categorize the first week of expenses
    - Create the first financial report
    - Review the first month's reconciliation
    - Complete the first accurate monthly close (Ultimate Win)

3. Payroll Journey

- Same-Day Win: Add first employee profile
- Journey Milestones:
    - Set up the first tax information
    - Enter the first employee's bank details
    - Complete the first payroll run (Ultimate Win)

PUTTING YOUR JOURNEY MAP INTO ACTION

In this chapter, we explored how to keep users motivated on their journey to the Ultimate Win by mapping out meaningful victories along the way. Like Locke's research showed, seeing clear evidence of progress—even small wins—dramatically increases motivation and persistence.

We start with a Same-Day Win that validates the user's decision. Then, we map out five to eight key wins that build toward the Ultimate Win. Together, these form an Onboarding Success Roadmap that maintains momentum throughout the implementation journey.

In the next chapter, we'll put these concepts into action with the Reverse Journey Map Exercise. Working with your onboarding team, you'll map out your specific wins journey by starting with your Ultimate Win and working backward to identify the key milestones that lead there. This collaborative exercise will help align your team around a clear onboarding path and ensure every milestone moves users closer to experiencing your product's core value.

EUREKA ACTION ITEMS: MAPPING YOUR ONBOARDING WINS

Start mapping out your onboarding journey wins with these steps:

1. List Potential Wins

- What are all the meaningful victories users experience?
- Which moments give users confidence they're succeeding?
- What achievements do they share with their team?
- When do they first experience your product's value?
- Use the format: [ACTION] the first [THING]

2. Structure Your Journey

- From your list of wins, identify your Same-Day Win.
- Map five to eight key wins between your Same-Day Win and Ultimate Win.
- Make sure each win builds toward your Ultimate Win.

3. Validate Your Journey

- Talk to successful customers about their early wins.

- Review product analytics for common progression patterns.
- Workshop with your customer-facing teams.

We'll do a comprehensive journey mapping exercise in the next chapter, but these initial steps will help you start thinking about your key wins.

12

The Reverse Journey Map Exercise

~~~

The best way to predict the future is to create it.

Peter Drucker

~~~

"How do we** actually identify the right wins for our product?"
The question comes up in nearly every onboarding workshop
I run. While the concepts of Ultimate Wins and onboarding milestones
that we covered in Chapters 10 and 11 make sense, translating them
into actionable steps for a specific product can feel overwhelming. The
Reverse Journey Map Exercise offers a structured solution.

By working backward from success, teams create a clear, achievable
path that guides users to their desired outcomes. You can find templates, facilitation guides, and recordings of this workshop in action
at *eurekabonus.com*.

THE REVERSE JOURNEY
MAP EXERCISE

The Reverse Journey Map Exercise is a collaborative workshop that
helps teams identify and document:

- Their product's Ultimate Win for each user journey—the key milestone where users first experience core value

- The Same-Day Win that builds initial momentum and confidence

- Strategic wins between these endpoints that maintain engagement

- Clear metrics to measure progress and moments worth celebrating

OBJECTIVE

The Reverse Journey Map Exercise helps teams identify and align on:

- Clear definitions of Ultimate Wins for each user journey

- Strategic milestones that mark meaningful progress

- Achievable Same-Day Wins that build early momentum

- Dependencies between different user journeys

- Metrics that matter for measuring onboarding success

By the end of this exercise, your team will have a shared understanding of your onboarding journey and a prioritized roadmap of wins that lead users to success.

A NOTE ABOUT USER RESEARCH

This exercise relies heavily on insights from customer interactions. The most valuable input comes from team members who regularly engage with users through:

- Sales conversations and demos
- Customer onboarding calls
- Support interactions
- User research sessions

Before running the workshop, gather:

- Recordings of 3–5 recent sales calls
- Notes from successful onboarding sessions
- Common challenges documented by support
- Success stories from customer champions

For self-serve products without direct customer contact, conduct user interviews first. You'll find interview guides and templates at *eurekabonus.com.*

THE SETUP

Time needed: 60–90 minutes

Participants: Cross-functional team of 4–8 people from:

- Product
- Marketing
- Sales
- Customer Success
- Support

Materials needed:

- Whiteboard or digital whiteboard
- Square sticky notes (different colors)
- Sharpies or markers
- Timer
- Completed User Success Canvas(es)

You can download templates for FigJam and Miro at *eurekabonus.com*.

FACILITATION TIPS

The success of this exercise depends not just on following the steps but on how well you facilitate the discussion and manage team dynamics. These tips will help you navigate common challenges and keep the exercise productive and engaging.

- **Stay User-Focused**: Regularly ask, "Is this a win for users?" and challenge any metrics that don't represent user value.

- **Maintain Momentum**: Use strict timeboxes, take breaks between sections, and keep discussions focused on one win at a time.

- **Navigate Challenges**: When stuck, return to the user experience first. Break down complex wins and prioritize based on user value.

- **Handle Multiple Journeys**: Use color coding to distinguish different journeys, identify shared wins, and maintain a master timeline showing all paths.

REVIEW SESSION: UNDERSTANDING OUR STARTING POINT (5–10 MINUTES)

Before diving into journey mapping, review the User Success Canvas(es) completed in Chapter 9 with your onboarding team. This ensures everyone understands the different user journeys we'll be mapping.

Have the team review each canvas, walking through:

- The Jobs-to-be-Done statement
- The three components of success
- Key forces enabling and blocking progress

Together, select your product's primary customer job as your starting point. This should be the most important journey that represents your core user's path to success.

With our foundation established, we're ready to begin the journey-mapping process. This exercise will consist of five distinct parts, each building on the previous one to create a comprehensive view of your onboarding journey.

THE EXERCISE

Step 1: Select Onboarding Journey Wins (15 minutes)

The first step is identifying your product's Ultimate Win—the crucial milestone where users first experience your core value. You'll do this through brainstorming, voting, and discussion.

1. Give each participant sticky notes and a marker.

2. Set a timer for 5–10 minutes. Ask participants to write down as many potential wins as possible:

    ○ Any achievement that moves users toward success
    ○ Use the format: "[ACTION] first [THING/TIME]"
    ○ One win per sticky note
    ○ Don't worry about organizing or filtering yet

3. Have everyone post their notes randomly on the board.

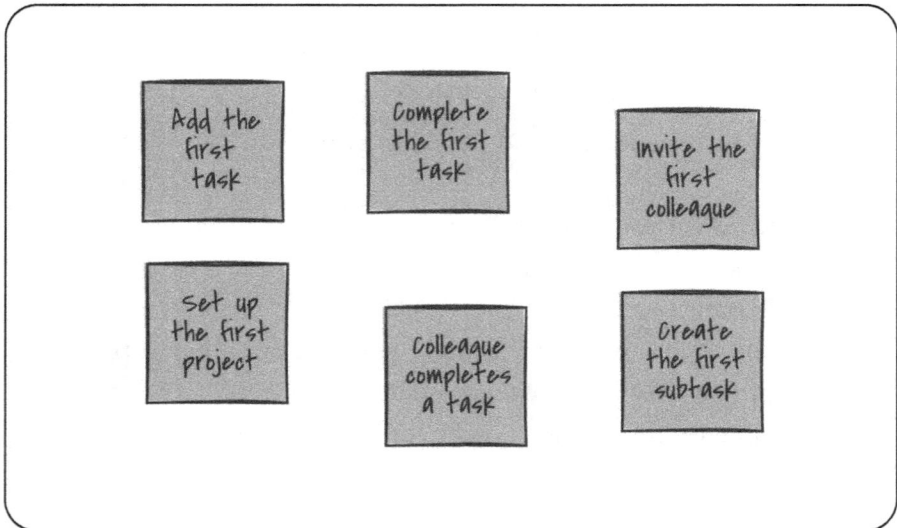

4. As a group, identify and remove duplicate wins.

5. Review the five characteristics of an Ultimate Win:

    ○ A genuine victory from the user's perspective
    ○ An early indicator of deeper transformation
    ○ Something trackable and measurable
    ○ Repeatable and habit-forming
    ○ Simple enough to achieve

6. Give each participant 3 voting dots and explain the voting rules:

    ○ Place dots on wins that you think represent the Ultimate Win
    ○ Can put multiple dots on a single sticky note

- ○ Can vote on your own ideas
- ○ If an idea isn't clear, skip it (no asking for clarification)
- ○ Must use all dots within the time limit
- ○ No discussion during the voting

7. Set a timer for 5 minutes to begin silent voting on potential Ultimate Win.

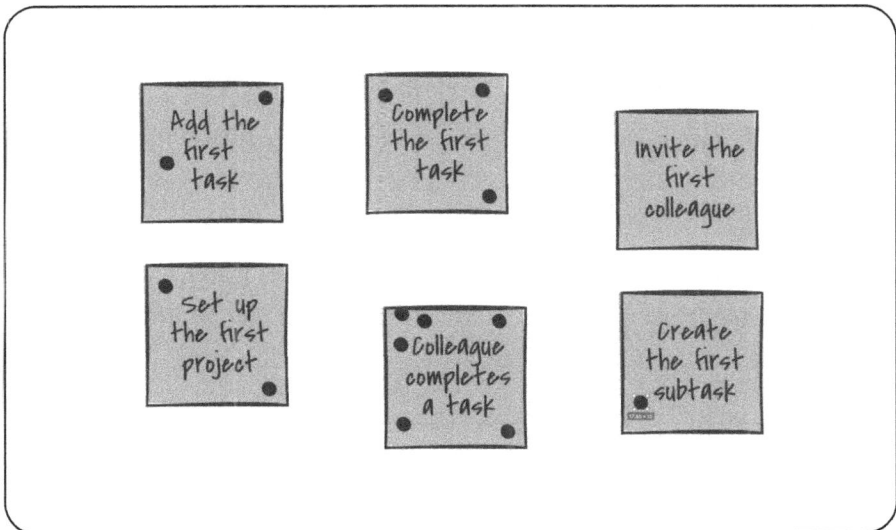

8. Facilitate a ten-minute team discussion:

- ○ Customer Success: Share stories of when customers first felt successful.
- ○ Sales: Describe what prospects say they want to achieve.
- ○ Marketing: Present data on what messaging resonates.
- ○ Product: Provide insights from usage analytics.
- ○ Review any available qualitative and quantitative data.
- ○ Challenge any wins that might be "selfish metrics."
- ○ Validate against the five characteristics.

PRO TIP: FOCUS ON USER VALUE

During voting and discussion of the Ultimate Win, watch out for "selfish metrics" that benefit the company but not users. Guide the conversation by asking:

- *"Would users celebrate achieving this?"*

- *"Does this represent genuine value for them?"*

- *"Is this a meaningful milestone in their transformation?"*

If the team gets stuck debating metrics, redirect focus to the user experience first. Metrics can be refined once you've aligned on the core user victory.

Step 2: Map the Wins

Now that we have our Ultimate Win, we'll map the journey that leads users there. We'll identify the key wins that guide users from signup to success.

1. Prep the Timeline (2 minutes)

- Draw numbers 1–8 on the board horizontally.
- Place the Ultimate Win at position 8.
- Leave space above/below each number for sticky notes.

2. Individual Journey Mapping (10 minutes)

Give each participant 7 sticky notes and have them silently map their version of the journey wins. Why 7? We want to focus on major milestones rather than getting lost in details—technical teams especially tend to document every small step. Ask participants to:

- Start with a Same-Day Win achievable in the first session

- Use the format "[ACTION] first [THING]" for each win

- Work forward chronologically toward the Ultimate Win

- Focus only on significant victories that move users forward

3. Place and Compare (15 minutes)

Have all participants bring their sticky notes to the board, placing similar steps above and below each other under the corresponding numbers. This vertical alignment makes it easy to compare how different team members envisioned each step of the journey.

Have a discussion as a group about:

- Patterns where multiple people identified the same wins

- Places where teams use different terms for the same milestone

- Gaps in the journey where "something magical happens" without clear steps

- Strong agreement on early wins, which often indicates a natural Same-Day Win

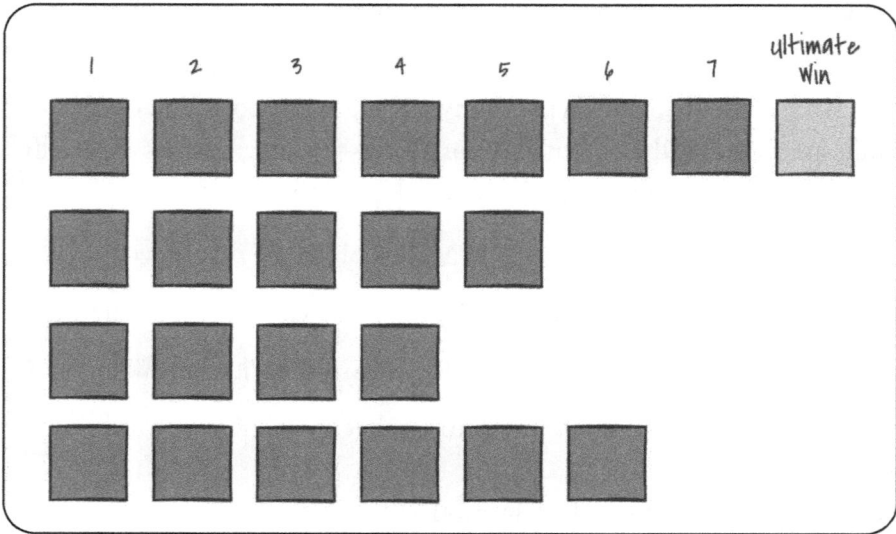

4. Finalize Journey (5 minutes)

As a group, select the final version of each win that will make up your onboarding journey. Review the complete sequence to ensure there's a clear progression from one win to the next.

PRO TIP: DESIGN FOR PROGRESSIVE SUCCESS

When mapping wins between Same-Day and Ultimate Win, think like a game designer creating progressive levels:

- *Early wins should come quickly (hours/days) to build confidence.*

- *Middle wins can be spaced further apart (days/weeks) as users gain momentum.*

- *Each win should be slightly more complex than the last but never overwhelming.*

- *If you spot a large gap between wins, that's often where users drop off.*

Remember: A good journey map feels like climbing stairs, not scaling a wall. Each step should be challenging but achievable.

Step 3: Repeat for Additional Journeys

Most B2B products serve multiple user types—from individual contributors to team leads to executives. Each may need their own journey to success. For each additional User Success Canvas:

1. Take a quick break (5–10 minutes).

2. Clear the board, but keep the previous journey visible for reference.

3. Repeat Steps 1 and 2 with the new user type in mind.

As you map these additional journeys, look for:

- Wins that could serve multiple user types
- Places where journeys intersect or depend on each other
- Natural order if users pursue multiple paths
- Opportunities to reuse or adapt existing wins

This helps create a comprehensive view of how different users progress through your product while identifying opportunities for shared resources and coordinated support.

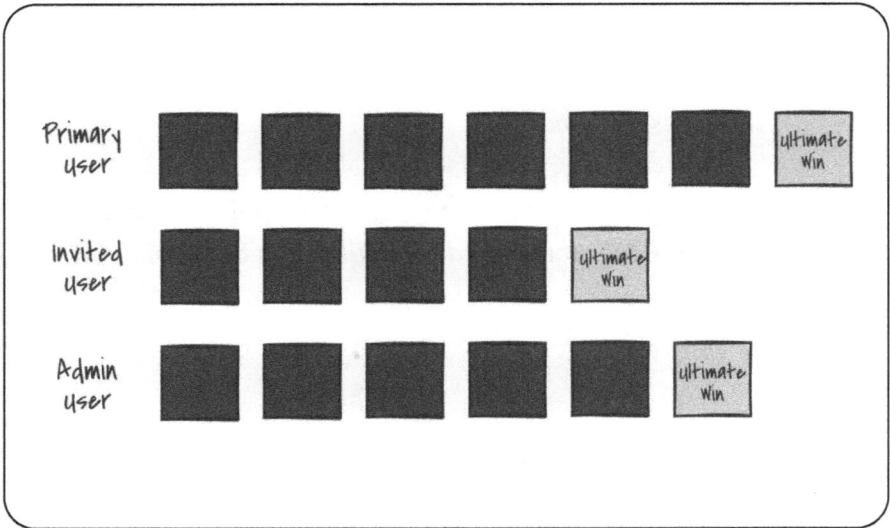

OUTPUT AND NEXT STEPS

While the exercise itself is valuable, its true impact comes from how you use its outputs. The journey maps you create will be blueprints for your onboarding experience, guiding everything from feature development to user communication.

The completed Reverse Journey Maps provide:

- Clear definition of Ultimate Wins for each user journey
- Engaging Same-Day Wins to build momentum
- Mapped journeys of progressive wins
- Understanding of journey interactions and dependencies
- Foundation for personalized onboarding paths

Use these insights to:

- Design targeted onboarding experiences for each journey
- Create adaptive progress indicators
- Plan celebration moments
- Set up analytics tracking for multiple paths
- Identify opportunities for journey optimization

EUREKA ACTION ITEMS: THE REVERSE JOURNEY MAP WORKSHOP

Ready to run your own Reverse Journey Map workshop? Download the template and facilitation guide from *eurekabonus.com*. Remember to:

1. Review all User Success Canvases beforehand.
2. Gather insights from customer-facing teams.
3. Prepare examples of potential wins for each journey.
4. Set clear expectations with participants.
5. Plan enough time for all journeys.

In the next step of the EUREKA Framework, we'll explore how to turn these journey maps into engaging onboarding experiences that guide users confidently toward their Ultimate Wins.

STEP 4

Keep New Users Engaged

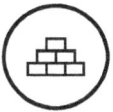

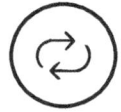

| Establish a Team | Understand Success | Reverse Journey Map | Keep Users Engaged | Apply Analyze and Repeat |

13

Friction Mapping and Onboarding Audit

~~~~

If I had an hour to solve a problem, I'd spend fifty-five minutes thinking about the problem and five minutes thinking about solutions.

Albert Einstein

~~~~

One of my life goals is to get my pilot's license. While watching training videos, I'm struck by how meticulously pilots map potential barriers before each flight—analyzing weather patterns, fuel requirements, and backup plans. A flight instructor's words stuck with me: "The key to safe flying isn't just knowing how to operate the aircraft—it's understanding exactly what could prevent you from reaching your destination."

Many companies approach onboarding like novice pilots, eager to take off without mapping the journey ahead. Just as pilots must consider weather conditions, mechanical limitations, and human factors, successful onboarding requires identifying all potential barriers users may encounter along their journey.

Building on our previous work with the Sailboat Exercise, User Success Canvas, and Reverse Journey Map, we'll now focus on the fourth step of the EUREKA Framework—keeping users engaged. This chapter introduces two exercises: the Friction Mapping Process to identify where users get stuck and the Three-Step Onboarding Audit to eliminate unnecessary steps and barriers in your current onboarding experience.

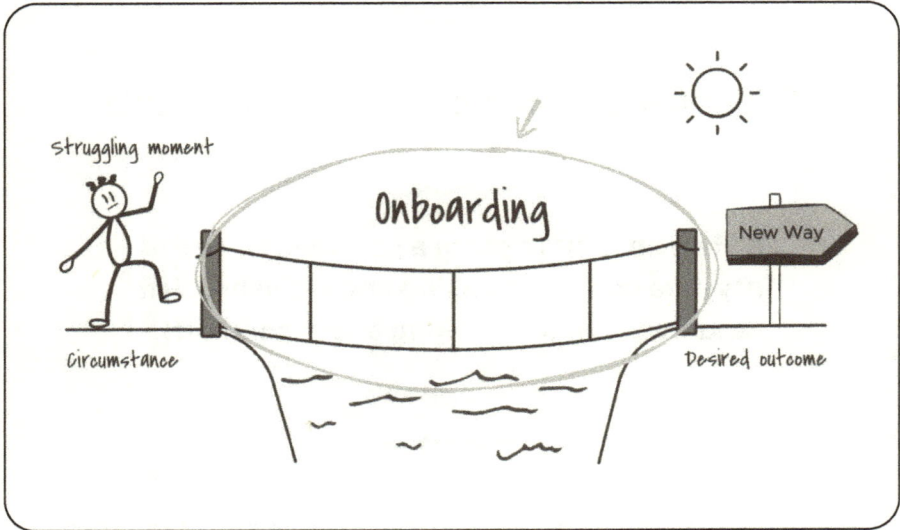

UNDERSTANDING ONBOARDING FRICTION

Most companies try to keep users engaged by adding more tactics to their onboarding—another product tour here, a training course there, maybe some customer success check-ins. While well-intentioned, this approach misses a crucial first step: understanding why users disengage in the first place.

Building on the three types of friction we explored in Chapter 2, let's see how functional barriers (like technical setup issues), social barriers (like team coordination), and emotional barriers (like confidence and trust) manifest throughout the user journey.

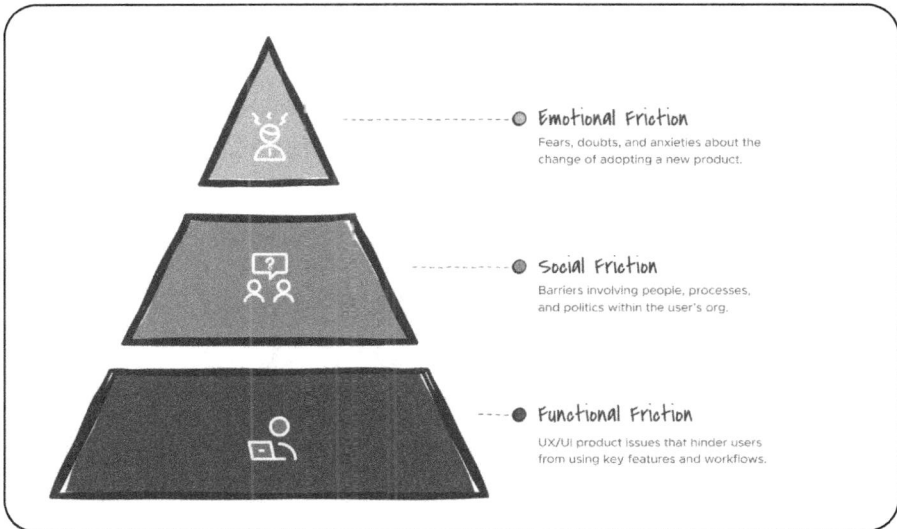

Let's see how these friction points manifest in a real example. Consider a project management tool's onboarding journey, specifically the milestone of "Add first team member:"

Functional Friction:

- Users struggle to find the team invitation feature.
- Email invitations fail to deliver.
- SSO and domain verification issues block progress.

Social Friction:

- Team members delay accepting invitations.
- Subscription seat limits create bottlenecks.

Emotional Friction:

- Users worry about setting proper permissions.
- Teams question data security.
- Teams resist adopting new tools.

Each milestone in your journey will have a unique combination of these friction points. Some steps might encounter primarily functional barriers, while others face more social or emotional challenges. The key is identifying which ones most significantly impact your users' progress toward their Ultimate Win.

BALANCING FRICTION WITH TIME-TO-VALUE

Before we dive into the exercises—a word of warning: Many teams fall into the trap of trying to eliminate all friction from their onboarding experience to reduce the time-to-value. While reducing unnecessary barriers is important, adding some friction in one area (like functional) can often help overcome larger barriers elsewhere (like social or emotional ones).

Take Superhuman, an email client: They used to require a live onboarding call before product access. While this creates functional friction (scheduling time, waiting for access), it strategically addresses deeper barriers:

- **Emotional Friction**: The personal connection builds trust and confidence when switching from familiar tools like Gmail.

- **Social Friction**: The human interaction creates accountability and investment in the learning process.

- **Functional Friction**: Direct guidance ensures proper understanding of keyboard shortcuts and workflows that drive long-term retention.

As you can see, intentional friction can lead to better outcomes when it serves a clear purpose. This insight reveals three key reasons why some friction is necessary in the onboarding:

1. True Mastery Takes Time

Rushing through the learning process for complex products often leads to frustration and abandonment. Consider the trade-off between immediate access and proper understanding.

2. Delayed Gratification Builds Value

The satisfaction of conquering a challenge creates deeper engagement than instant gratification. As we explored in Chapter 11, space your wins strategically—start with quick early victories (hours/days apart), then gradually increase the gap between achievements (weeks apart) to build sustained confidence.

3. Meaningful Habits Form Slowly

Products designed for long-term engagement benefit from a measured approach that builds confidence gradually. Weigh the cost of additional friction against the long-term benefits of proper adoption.

Let's put these principles into practice. The team-based workshops below will help you make intentional trade-off decisions about which friction points to preserve and which to eliminate. After all, these choices are too important for any one person to make alone.

THE FRICTION MAPPING EXERCISE

The Friction Mapping Exercise brings your team together to systematically identify and address barriers in your onboarding journey. Through collaborative mapping of functional, social, and emotional friction points, your team will:

- Surface hidden barriers across the entire user journey

- Create a shared understanding of key challenges

- Align on priority areas for improvement

- Build consensus on which friction points to tackle first

Before You Begin

Before running this workshop, gather:

- Recent support tickets and customer feedback

- Insights from sales and success teams

- Product usage data, if available

- Common user struggles and drop-off points

This preparation grounds discussions in real user experiences rather than assumptions.

PRO TIP: MAKE SESSION WATCHING A TEAM RITUAL

When Andrew Capland led growth at Wistia, his team transformed their onboarding through a simple ritual: Fullstory Fridays. Every week, the team gathered to watch twenty new users interact with their product for the first time.

"It was painful," Andrew recalls. "Users got lost, rage-clicked unclickable elements, and sometimes left right before their 'aha' moment. Once, we saw someone get three popups simultaneously."

But these sessions became a game changer. The team left each viewing with actionable improvements: bug fixes, copy changes, UI updates, and experiment ideas. While they used Fullstory, tools like Hotjar or LogRocket work just as well.

The key? Make session watching a regular team activity. Nothing beats seeing real users interact with your product—that's where the magic happens.

The Setup

Time needed: 30–45 minutes

Participants: Your cross-functional onboarding team (4–8 people)

Materials needed:

- Square sticky notes (different colors for different types of friction)
- Sharpies or markers
- Timer
- Your completed Reverse Journey Map
- Whiteboard or digital whiteboard
- Create three rows below your Reverse Journey Map:
  - Row 1: Functional friction (technical and usability barriers)
  - Row 2: Social friction (team and organizational barriers)
  - Row 3: Emotional friction (psychological and confidence barriers)

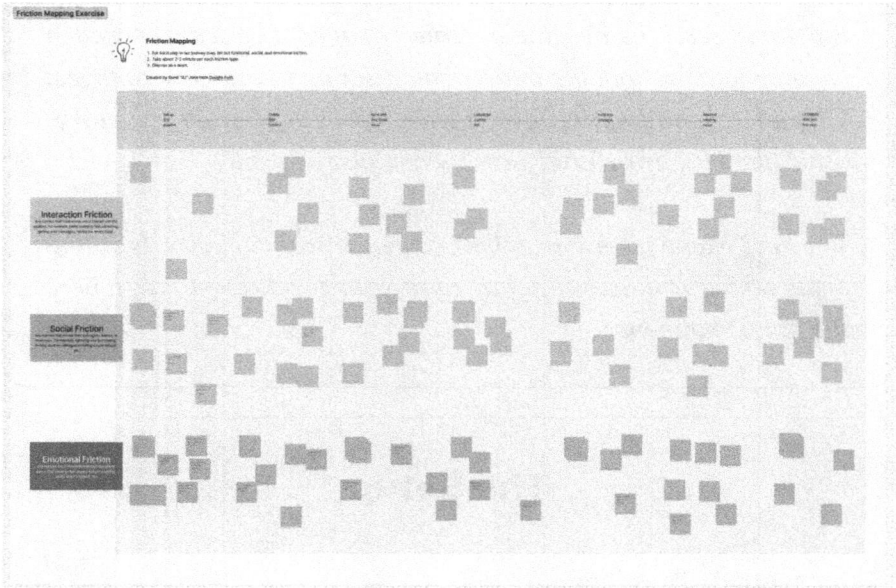

You can also use a template I've created for FigJam and Miro. You can access it at *eurekabonus.com*.

The Activity

1. Map Friction Points (30 minutes total)

Display your Onboarding Success Journey from Chapter 12 on a board or digital whiteboard. You'll map friction in three rounds:

Step 1: Map Functional Friction (10 minutes)

1. Give each team member sticky notes and markers.

2. Set a timer for 5 minutes.

3. Have everyone silently write down functional friction points they've observed along the journey.

4.   Write one friction point per sticky note.

5.   Ask them to consider:
   o   Technical barriers
   o   Usability issues
   o   Knowledge gaps
   o   Performance problems

Pause to share and discuss. Have each person place their sticky notes on the journey where the friction occurs.

Step 2: Map Social Friction (10 minutes)

1.   Repeat the same process with different colored sticky notes for social friction points.

2.   Focus on organizational and team friction:
   o   Stakeholder resistance
   o   Team coordination issues
   o   Resource constraints
   o   Process changes

Pause again to share and discuss. Add sticky notes to the journey.

Step 3: Map Emotional Friction (10 minutes)

1.   Repeat the same process with different colored sticky notes for emotional friction points.

2.   Consider psychological barriers:
   o   User fears and anxieties
   o   Confidence issues
   o   Trust concerns
   o   Change resistance

Share and discuss one final time. Complete the Fiction Map.

Step 4: Identify Patterns (15 minutes)

1. **Look across all friction points on your journey map.**
2. **Group similar friction into categories:**
    1. Functional patterns (technical issues, usability problems)
    2. Social patterns (team resistance, stakeholder challenges)
    3. Emotional patterns (user fears, confidence issues)
3. **Mark the most common friction points.**
4. **Note where multiple types of friction cluster around specific milestones.**
5. **Discuss patterns as a team:**
    1. Which friction points appear most frequently?
    2. Where do different types of friction overlap?
    3. Which milestones face the most barriers?
    4. What are the biggest things preventing users from reaching their Ultimate Win?
6. **Document and finalize the Fiction Map that clearly shows:**
    1. Journey milestones
    2. Friction points at each stage
    3. Common patterns and themes
    4. Priority areas to address

Understanding these patterns helps you choose the right combination of solutions. For example:

- Functional friction often responds well to in-product guides and UI improvements.

- Social friction might need educational content about change management.

- Emotional friction could require human touchpoints to build confidence.

Understanding these patterns helps you identify where users face the biggest barriers to success. In Step 5 of the EUREKA Framework, we'll explore how to turn these barriers into an action plan, using the three pillars of successful onboarding—in-product guides, educational content, and human touchpoints—to systematically overcome each type of friction.

Now that you've mapped where users encounter friction, the next step is evaluating your current onboarding experience. The Onboarding Flow Audit will help you identify which parts of your existing flow might be creating unnecessary friction.

THE THREE-STEP ONBOARDING AUDIT PROCESS

If you already have an existing onboarding flow—whether it's a series of product screens or a high-touch process with human interactions—you can conduct an optional Onboarding Flow Audit. This exercise helps evaluate your current onboarding experience against your Onboarding Success Journey to identify gaps and opportunities for improvement.

Objective

Through this collaborative workshop, your team will:

- Evaluate your current onboarding flow
- Identify unnecessary friction points
- Map opportunities for improvement
- Create alignment on priority changes

The Setup

This exercise works best when you have an existing onboarding flow (either product screens or a high-touch process) and a mix of product and customer-facing team members.

Time needed: 30–45 minutes

Participants: Your cross-functional onboarding team (4–8 people)

Materials needed:

- Whiteboard or digital whiteboard
- Sticky notes or voting dots (three colors for voting: green, yellow, red)
- Your current onboarding flow documentation
- Timer
- Your completed Reverse Journey Map

You can download templates for FigJam and Miro at *eurekabonus.com* to help facilitate this exercise.

Facilitation Tips

- Encourage healthy debate about what's essential.
- Listen to all perspectives (Product, Sales, Customer Success).
- Focus on data when available.
- Keep discussion centered on user value.

The Exercise

Step 1: Map Current Flow (15 minutes)

Assign someone from your onboarding team to take your Onboarding Success Journey and map out all the steps needed to complete each win. Include:

- Product screens and interactions
- Human touchpoints (like kickoff calls or training sessions)
- Email communications
- Documentation requirements
- Setup processes

Add screenshots of your in-product signup and onboarding flow. For in-person sessions, use sticky notes to represent each touchpoint. This makes it easy to visualize and reorganize the flow.

Step 2: Evaluate Steps (15 minutes)

1.   Give each participant a green, yellow, and red voting dot.

2.   Set a timer for 10 minutes.

3.   Have each person silently place a green, yellow, or red voting dot next to each step:

     o   Green: Essential steps that directly contribute to Ultimate Win
     o   Yellow: Steps that could be simplified or moved later after their Ultimate Win
     o   Red: Steps creating unnecessary friction

4.   Take 5 minutes to discuss steps with mixed votes.

Step 3: Finalize with Your Team (15 minutes)

This is where the fun begins! Each team member will have different perspectives on what's mission-critical versus unnecessary. Your product team might point to analytics showing certain steps cause significant drop-offs. Meanwhile, Sales might insist that collecting phone numbers during signup is crucial for their follow-up process, and Customer Success might advocate for keeping certain human touchpoints they know to prevent implementation failures.

These healthy debates reveal important tradeoffs between conversion optimization, sales needs, and customer success requirements. The goal isn't to eliminate all friction—it's to make intentional choices about which friction points are worth keeping.

1.   Review steps with mixed votes first.

2.   Have team members explain their reasoning.

3.   Focus discussion on:

     o   Does this step directly contribute to Ultimate Win?
     o   Could this step wait until after initial success?
     o   Is there overlap between touchpoints?
     o   Could we replace this with a scalable solution?

4. Make final decisions:

   ○ Remove red-marked steps creating unnecessary friction.

   ○ Move yellow-marked steps to the post-Ultimate Win phase.

   ○ Keep essential green steps in the initial flow.

5. Document decisions and next steps.

The audit process helps align your actual onboarding flow with the ideal journey you've mapped out. To get started, download the Onboarding Flow Audit template at *eurekabonus.com*.

TAKING YOUR NEXT STEPS

At this point, you have two powerful tools: your detailed map of friction points across your onboarding journey and an audit of your current onboarding experience. These tools give your team insights to:

- Remove unnecessary friction points
- Simplify complex steps
- Create clearer paths to value
- Plan systematic improvements

Before diving into specific solutions, in Chapter 14, we'll explore the behavioral psychology principles that drive successful onboarding—from reducing the cognitive load to building user confidence. Then, we'll examine how to implement these principles through The B2B Onboarding Toolkit in Chapter 15.

**EUREKA ACTION ITEMS:
MAPPING FRICTION AND
AUDITING YOUR ONBOARDING**

Ready to identify friction points in your onboarding journey?
Here's what to do:

1. Gather your onboarding team.

2. Map friction points using the Friction Mapping
   Exercise.

3. Look for patterns and common challenges.

4. Conduct the Three-Step Onboarding Flow Audit
   Process to evaluate your current experience.

Download the Friction Mapping and Flow Audit templates at
eurekabonus.com to get started.

In Chapter 16, we'll bring everything together in the Friction-to-Action Workshop, where you'll transform these insights into an actionable roadmap for improvement.

14

Behavior Psychology Principles for Successful Onboarding

～～～

If you pick the right small behavior and sequence it right, then you won't have to motivate yourself. It will just happen naturally, like a good seed planted in a good spot.

Dr. BJ Fogg, Behavior Scientist and Adjunct Professor at Stanford University

～～～

Every January, millions of people make the same resolution: "I'll wake up at 5 a.m. to go to the gym." Yet by February, most have abandoned this goal. It's not because they lack motivation or don't understand the benefits of exercise. The real barrier? Cognitive load—the mental effort required to change behavior.

Dr. BJ Fogg's research on behavior change reveals why this happens. When we're tired or stressed (low ability), even simple tasks like finding gym clothes in the dark feel overwhelming. But when we reduce the cognitive load by preparing the night before—laying out clothes, packing our bags, and planning our route—we're much more likely to succeed. The same task becomes easier because we've reduced the mental effort required.

The friction mapping process and onboarding audit in Chapter 13 revealed similar patterns in B2B software adoption. Fundamental truths about human behavior lie beneath every technical hurdle, stakeholder resistance, and implementation challenge. Users who face a complex setup process struggle not just with the number of steps but with the cognitive load each step demands. Teams that resist adoption battle not just with training requirements but with deep psychological barriers to changing established habits.

In this chapter, I'll share seven behavioral psychology principles that will help you reduce cognitive load and build lasting engagement in your onboarding experience:

1. Miller's Law: Break down complex steps

2. Hick's Law: Limit choices for new users

3. Progressive Overload: Start small to build habits

4. Goal Gradient Effect: Show progress to motivate

5. Zeigarnik Effect: Use incomplete states to drive action

6. IKEA Effect: Let users customize to build investment

7. Mirror Neurons: Demonstrate instead of explaining

These principles provide guidelines for implementing the three pillars of successful onboarding we'll explore in Chapter 15—in-product guides, educational content, and human touchpoints.

Let's start by examining how to make powerful first impressions that build user confidence rather than create confusion.

1. MILLER'S LAW: BREAK DOWN COMPLEX ONBOARDING STEPS

Psychologist George Miller discovered that humans can only process about seven pieces of information at once—beyond that, our ability to understand and remember drastically decreases. Smart businesses design around this limitation. Fast food restaurants, for example, offer five to seven pre-packaged meals that simplify multiple decisions into a single choice rather than overwhelming customers with endless combinations.

Unlike ordering a pre-packaged meal, B2B products face a bigger challenge: They require multiple setup steps before users see any value. A typical CRM implementation, for instance, requires users to configure the pipeline, customize fields, set up automation, and import data—all complex tasks that must be completed before seeing the product's full value.

The takeaway: Break complex processes into focused phases that create momentum through quick wins.

Instead of one lengthy process, guide users through three manageable phases: basic setup ("Get Your First Deal Tracked"), team workflows ("Customize Your Pipeline"), and advanced features ("Automate Your Process"). Each phase should contain no more than four to five steps, letting users experience success before tackling the next challenge.

For your product, ask yourself:

- Which setup processes exceed seven steps?
- How can you break them into focused phases?
- Where are the natural break points for celebrating wins?

The answers to these questions will help you avoid overwhelming users with too much complexity. But even with simplified steps, users face another challenge: too many choices at once.

2. HICK'S LAW: LIMIT CHOICES FOR NEW USERS

Hick's Law states that decision time increases with the number of choices available. Apple's iPhone setup demonstrates this principle well—instead of overwhelming users with hundreds of settings, it presents just two to three essential choices per screen. Users make simple decisions about language, WiFi, and security before accessing advanced options in Settings. While consumer products like the iPhone embrace simplicity, B2B software often falls into the opposite trap: showcasing capability through options.

The temptation in B2B software is to showcase capability through options. Product teams often pride themselves on flexibility—multiple ways to visualize data, various automation rules, and countless integration options. However, this flexibility often paralyzes new users who lack context for these choices. Even simple decisions like choosing between "list view" or "board view" create unnecessary friction when presented too early.

The takeaway: Limit options to essential actions and hide advanced features until later.

Zoom demonstrates this perfectly with one prominent "Start Meeting" button, keeping advanced options tucked away in menus. Users succeed quickly with core features and then naturally discover additional capabilities as their needs grow.

For your product, ask yourself:

- Which features can wait until after initial success?
- How can you simplify early decisions?
- Where can you hide advanced options without removing them completely?

Understanding these decision points helps create clear pathways to value instead of overwhelming users with possibilities. But even with limited choices, users need a structured path to master complex features.

3. PROGRESSIVE OVERLOAD: START SMALL TO BUILD HABITS

Athletes build strength by gradually increasing weights over time, not by attempting their maximum lift on day one. The same principle applies to product adoption—**users build lasting habits through incremental steps and small wins**.

Video game designers have mastered this concept. *Super Mario*'s World 1-1 teaches basic movements through simple challenges before introducing tougher enemies and more challenging obstacles. Each level builds on previous skills while adding complexity. Similarly, Duolingo starts users with basic phrases like "the cat" before introducing daily streaks and competitions, making the commitment to practice grow naturally from tiny initial steps.

The takeaway: Organize experiences from easiest to hardest.

Start users with quick wins like profile customization, then gradually introduce individual workflows, team collaboration, and, finally, advanced features. Apply this progression everywhere—from onboarding checklists to educational content and product tours—letting each small victory build confidence for the next challenge.

For your product, ask yourself:

- What's the simplest valuable action users can take?

- How can you sequence features by complexity?

- Where can you celebrate small wins?

Small wins can transform new users into power users, but only if they stay motivated throughout their journey. This leads us to our next principle, which is maintaining momentum.

4. GOAL GRADIENT EFFECT: SHOW PROGRESS TO MOTIVATE NEW USERS

The Goal Gradient Effect reveals that humans and animals increase their effort as they near a goal—we run faster toward the finish lines and study harder before deadlines. Smart products tap into this psychological principle to maintain user momentum.

HubSpot demonstrates this principle by showing progress indicators and celebrating milestones throughout the onboarding journey. Instead of focusing on what's left, they show users they're "60 percent done with setup" and celebrate key wins like "first email sent" or "first lead captured." These multiple finish lines maintain momentum through long implementation processes.

The takeaway: Make progress visible and celebrate completion.

Break lengthy implementations into clear phases, each with its own finish line and celebration moment. While Miller's Law focuses on cognitive capacity (breaking down complex steps to avoid overwhelming users), the Goal Gradient Effect is about motivation and effort—showing users how close they are to success makes them more likely to increase their effort and complete the journey. Use progress bars, milestone celebrations, and clear "distance to goal" indicators to tap into users' natural tendency to sprint toward the finish lines.

For your product, ask yourself:

- Where can you show progress visually?
- Which milestones deserve celebration?
- How can you make the finish line feel closer?

Progress indicators can transform long journeys into achievable sprints. But sometimes, leaving tasks incomplete can create an even stronger pull.

5. ZEIGARNIK EFFECT: USE INCOMPLETE STATES TO DRIVE ACTION

Our brains hate leaving tasks unfinished. Psychologist Bluma Zeigarnik found that waiters remembered open orders better than completed ones—we're naturally driven to close these mental loops.

B2B products can create similar momentum by making incomplete tasks visible. For example, showing an onboarding checklist with one item already checked—"Set up your account"—creates two effects: Users feel immediate progress and feel drawn to complete the remaining items.

The takeaway: Make incomplete states visible but achievable. Show users their unfinished tasks while keeping them feeling within reach.

For your product, ask yourself:

- Which tasks can create helpful open loops?
- How can you make incomplete states visible?
- Where can you show partial progress?

Beyond maintaining momentum, we need ways to create deeper product investment through user participation.

6. IKEA EFFECT: LET USERS CUSTOMIZE TO BUILD INVESTMENT

People value what they help create. Despite its frustrations, the satisfaction of building IKEA furniture creates a stronger emotional connection than buying pre-assembled pieces.

B2B products can tap into this same psychology by letting users customize your product to match their needs. Notion exemplifies this approach by starting users with a blank canvas and basic building blocks. Instead of pre-built solutions, users gradually create their perfect workspace through customization. Each template modified and dashboard configured deepens their investment in the product.

As Yael Mark, Founder of Behavioral PM, notes: "We want to feel involved, but not exhausted. Identify the steps that require minimal effort yet have a meaningful impact—like choosing a shoelace color in a custom sneaker design rather than the entire pattern." This balance between customization and effort is crucial for maintaining user engagement without overwhelming them.

The takeaway: Give users opportunities to customize and create.

Guide them through personalizing their workspace, but let them feel ownership over the final result.

For your product, ask yourself:

- Which elements can users customize early?
- Where can you replace pre-built with user-built?
- How can you celebrate user customizations?

While customization builds investment, users still need to know how to use what they've built.

7. MIRROR NEURONS: DEMONSTRATE INSTEAD OF EXPLAINING

Research by neuroscientists Giacomo Rizzolatti and Laila Craighero at the University of Parma revealed that our brains contain "mirror neurons" that fire both when we perform an action and when we watch others perform the same action. That's why demonstration consistently outperforms written instruction.

Pitch's presentation software leverages this brilliantly with quick GIFs that show users how to customize slides or add animations. Instead of reading documentation, users watch and immediately replicate the actions. Short, looping demonstrations remove the friction of traditional video tutorials—no need to hit play or rewind to catch details.

The takeaway: Show, don't tell. Replace lengthy documentation with brief visual demonstrations that users can immediately mimic.

For your product, ask yourself:

- Which features need visual demonstration?
- Where can GIFs replace written instructions?
- How can you make complex actions feel achievable?

Visual demonstrations accelerate learning by activating users' natural ability to learn through observation.

PUTTING THESE PRINCIPLES INTO PRACTICE

In the next chapter, we'll explore applying these behavioral principles through the three pillars of successful onboarding introduced in Chapter

3: in-product guides, educational content, and human touchpoints. Together, these three pillars create a comprehensive system for guiding users to success. Let's examine how to implement these principles through each pillar to create an engaging onboarding experience that drives lasting adoption.

15

The B2B Onboarding Toolkit

~~~

**For every problem, there's a tool. The
trick is finding the right one.**

My Dad

~~~

One summer, I tried fixing our kitchen sink without the right
tools. Armed with just a wrench and determination, I spent hours
struggling with a stubborn pipe, only making the problem worse. When
my dad arrived with his plumbing toolkit, he fixed it in minutes.

"For every problem, there's a tool," he said, showing me specialized
wrenches I didn't know existed. "The trick is finding the right one."

My dad's advice has stuck with me throughout my career, especially
when helping companies improve their onboarding. Just as my failed
sink repair showed the importance of having the right tools, successful
onboarding requires a well-equipped toolkit—one that helps users navigate complexity, build confidence, and achieve lasting transformation.

In Chapter 3, we introduced the three pillars of B2B onboarding:
in-product guides, educational content, and human interaction. Now
that we've explored the behavioral psychology principles that drive
successful onboarding, it's time to examine these tactics in detail.

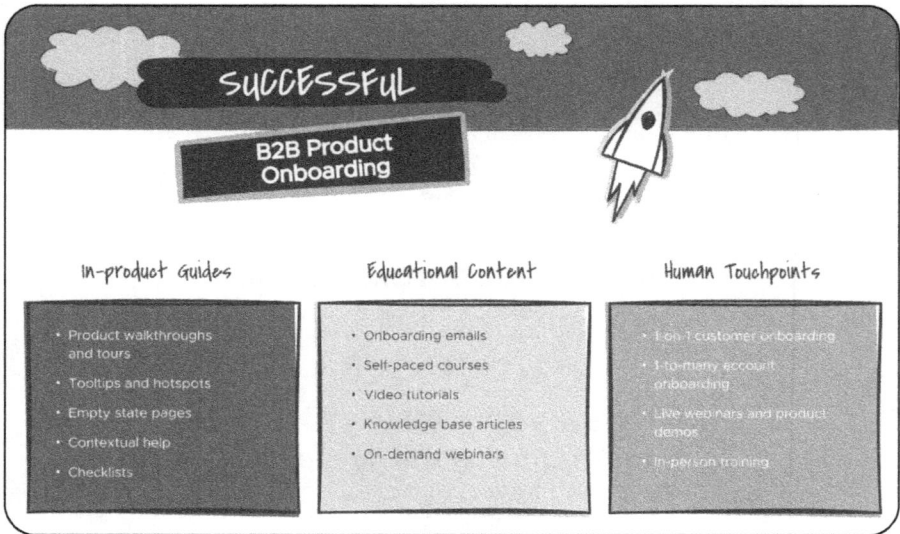

Think of this chapter as opening up your onboarding toolbox, examining each instrument's purpose, and understanding when to use each one. It serves as your reference guide to these tools. Like my dad's plumbing kit, not every tool will be right for every situation. Your product's complexity, market position, and team resources will determine the best approaches. Your onboarding team, with their deep understanding of your customers and product, will know better than I ever could which tools fit your specific needs.

So, feel free to skip over sections that may not apply to your product. Take what resonates, adapt what's promising, and ignore what doesn't fit. Consider this chapter a catalyst for ideas you'll use in the next chapter's Friction-to-Action Workshop, where you'll design targeted solutions for your specific challenges.

For each onboarding tactic, I'll share when to use it, examples from companies doing it well, best practices, and further resources. Let's start with the foundation of any product onboarding experience: in-product guides.

PILLAR 1: IN-PRODUCT GUIDES

~~~~~

Overload, clutter, and confusion are not attributes of information, they are failures of design.

Golden Krishna, Designer and Author of
The Best Interface is No Interface

~~~~~

In-product guides are UI elements that provide contextual help when and where users need it. Like street signs in a city, they help users navigate your product confidently. But like any good guide, they must know when to speak up and when to step back.

In this section, we'll explore three types of in-product guides:

- **Annotated guides** that direct attention to specific elements
- **Embedded guides** that provide a broader context
- **Dedicated guides** that create immersive learning moments

Annotated Guides

Annotated guides are UI elements that highlight and explain specific features where users need them. They come in two forms: tooltips and hotspots.

1. Tooltips: Guide Sequential Actions

Tooltips are floating messages near a feature, button, or section of your product. While individual tooltips explain specific elements, you can combine multiple tooltips to create guided product tours.

Example: Canva uses a four-step tooltip tour to teach new users essential design actions: customize elements, update text, add images, and download designs.

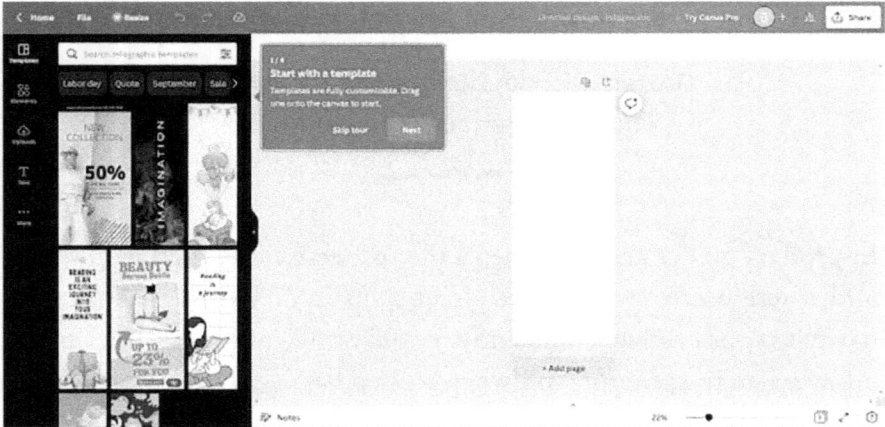

When To Use Tooltips:

Tooltips work best for linear processes requiring step-by-step guidance to complete tasks. Avoid using tooltips for complex features that require deeper understanding—in these cases, combine them with documentation or human support. Think of tooltips as training wheels: helpful for getting started but not meant for long-term use.

Tooltip Best Practices:

1. **Keep It Task-Focused**: Instead of explaining features, guide users through meaningful tasks. Instead of saying, "Here's the template gallery," say, "Let's create our first document using a template."

2. **Make Tours Less than Five Steps:** If your tour exceeds five steps, brace yourself for sharply declining completion

rates. More than half of users lose attention to tours beyond five steps. See Appendix II for more detailed product tour completion rate benchmarks.

3. **Show Progress**: As we explored with the Goal Gradient Effect in Chapter 14, users increase their effort as they near a goal. To tap into this psychology, include clear progress indicators like "Step 2 of 4."

4. **Break Complex Processes into Mini-Tours**: Following Miller's Law from Chapter 14, break lengthy processes into focused sequences. A project management tool might start with "Create Your First Project" (3 steps), then progress to "Invite Team Members" (2 steps) once users are comfortable.

5. **Give Users Control**: Click-triggered tours have completion rates of nearly 60 percent compared to 30 percent for auto-triggered ones. So, avoid automatically showing a product tour. See Appendix II for more product tour benchmarks.

2. Hotspots: Highlight Key Features

Hotspots are subtle visual indicators (like glowing orbs) that users can hover over or click to learn about features. Unlike tooltips that appear automatically, hotspots wait for user interaction, making them perfect for drawing attention to new or advanced features while letting users explore at their own pace.

Example: Hotjar demonstrated the power of hotspots through A/B testing when introducing their "relevance selector" feature. They tested the same content in a tooltip tour and a hotspot format. The hotspot not only drove higher feature adoption (8.16 percent vs. 6.5 percent with tooltips, a 26 percent improvement) but also led to a 99 percent increase in users saving relevant recordings for later review. This success showed how hotspots' non-intrusive nature can increase engagement by letting users discover features at their own pace.

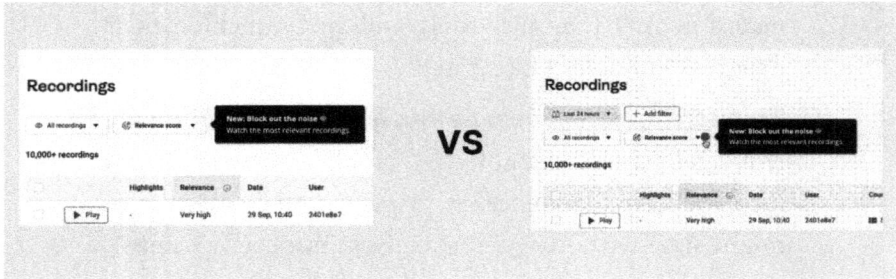

When to Use Hotspots:

Hotspots excel in feature-rich interfaces where users benefit from self-paced discovery. They're ideal for highlighting advanced features or new capabilities without interrupting the user's workflow.

Hotspot Best Practices:

1. **Keep It Focused and Visible**: Limit to four to five hotspots per page and ensure they stand out visually. Like well-placed street signs, they should be noticeable without being distracting.

2. **Write Value-Focused Text**: Instead of "Advanced Filters Available," try "Segment data to uncover hidden trends." Show users what they can achieve, not just what features exist.

3. **Space for Discovery**: Position hotspots strategically across the interface, giving each one room to breathe. When highlighting multiple features, ensure they don't compete for attention.

Embedded: Provide Broad Guidance in Context

Embedded guides are UI components that create dedicated spaces for learning and progress tracking. They come in three forms: modals for focused attention, checklists for systematic progress, and slideouts for detailed help.

1. Modals or Popup Windows: Capture Full Attention

Modals are overlay windows that interrupt the workflow at critical moments, such as user segmentation, preference collection, or major announcements.

Example: Eventbrite's welcome modal segments users with two clear paths: "Discover an event" for attendees or "Host an event" for organizers. This simple choice personalizes the entire experience.

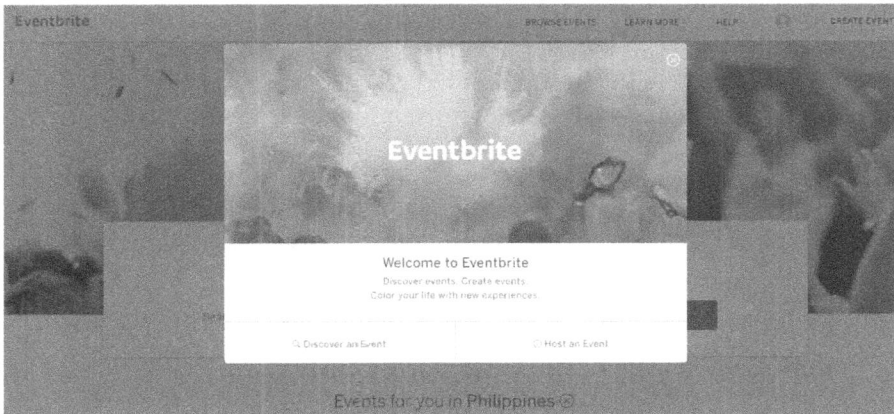

When to Use Modals:

Use modals for truly critical moments that deserve users' full attention, such as initial product setup, key preference selection, or major feature announcements. They're the digital equivalent of a "Stop and Read" sign, so use them sparingly.

Modal Best Practices:

1. **Make Next Steps Crystal Clear**: Every modal should have one obvious action users need to take. Make that action stand out with prominent call-to-action buttons and clear, action-oriented text like "Start Your First Project" instead of vague labels like "Continue" or "Next."

2. **Show Clear Progress**: For multi-step modals, include progress indicators ("Step 2 of 3") and consider adding visual elements like GIFs or screenshots to clarify actions.

2. Checklists: Drive Systematic Progress

Checklists combine psychological motivation with practical guidance, tapping into our natural desire for completion. They break complex processes into achievable steps while showing clear progress.

Example: Appcues splits its onboarding into focused phases. The Phase 1 checklist covers installation only. Once completed, the Phase 2 checklist appears with three items: build a flow, track an event, and create a goal. This keeps users focused on immediate tasks while maintaining momentum.

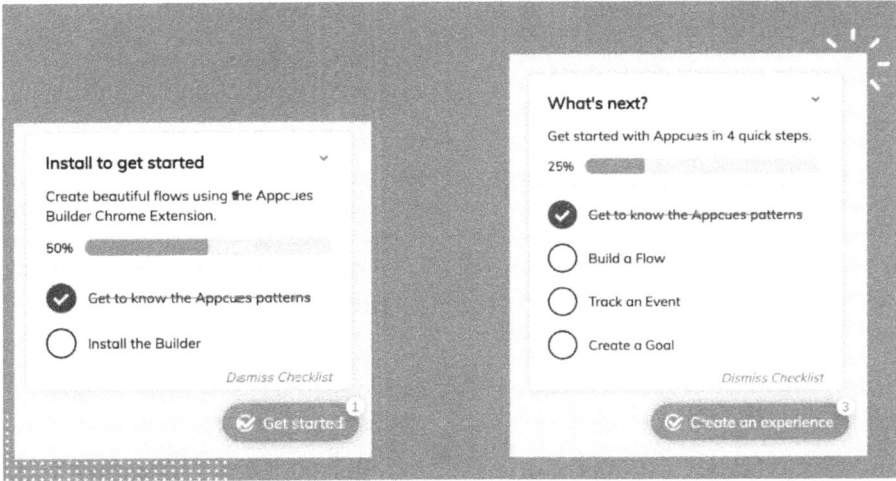

When to Use Checklists:

Checklists work best when users need to complete a series of related tasks to reach a meaningful milestone. They're particularly effective during initial setup, feature implementation, or any process that requires multiple steps to achieve value.

Checklist Best Practices:

1. **Start with Momentum**: Begin with one item already checked off (like "Create Account") to tap into the Zeigarnik Effect from Chapter 14. That first checkmark motivates users to keep going.

2. **Keep Lists Focused**: Limit each checklist to five items maximum and break longer processes into multiple focused lists. Each item should drive clear action, not just provide information.

3. **Show Clear Progress**: Help users see their advancement with progress indicators ("3 of 5 completed") and cele-

brate meaningful completions. Consider personalizing checklists based on user roles—technical steps for admins and basic usage for end users.

3. Slideouts: Provide Detailed Help

Slideouts are panels that appear from the screen's edge to show documentation, offer support, or preview content without losing context.

Example: Airtable's slideout celebrates "Great start with Cobuilder!" and offers an easy-to-copy invite link to "help build out your app." It's simple, contextual, and collaboration-focused.

When to Use Slideouts:

Slideouts shine when offering contextual resources without interrupting workflow. Use them to share webinar invites, downloadable templates, or demo scheduling options at key moments in the user's journey. They provide a middle ground between tooltips' quick hints and modals' full-attention moments.

Slideout Best Practices:

1. **Stay Consistent and Accessible**: Position slideouts consistently (stick to one side of the screen) and make them easy to minimize or dismiss. Users should always feel in control of their workspace.

2. **Keep Context in Mind**: Trigger slideouts based on user actions or context, not randomly. They should feel like a natural extension of the user's task, not an interruption.

Dedicated: Capture Data and Build Motivation

Dedicated guides are full-page experiences that command complete attention for key moments in the user journey.

1. Signup Flows: Personalize the Journey

Signup flows collect essential information to personalize the user's journey. They gather preferences and goals to deliver a tailored experience from the start.

Example: Notion asks about roles, team size, and use cases to customize everything from templates to feature highlights.

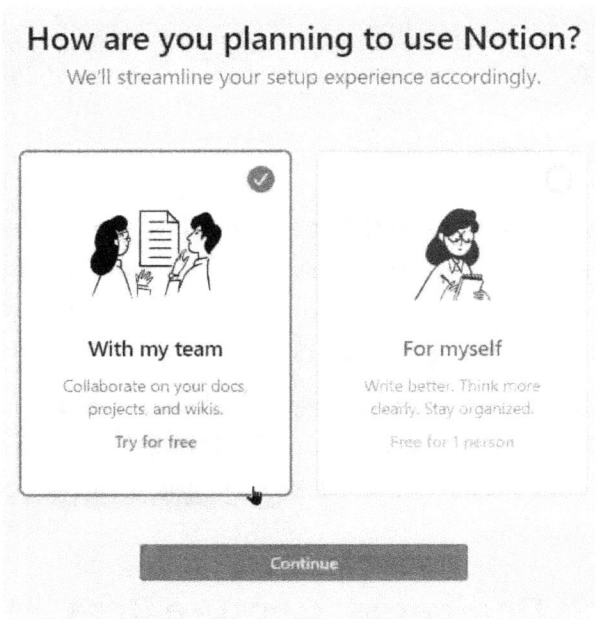

How are you planning to use Notion?

We'll streamline your setup experience accordingly.

With my team

Collaborate on your docs, projects, and wikis.

Try for free

For myself

Write better. Think more clearly. Stay organized.

Free for 1 person

Continue

When to Use Signup Flows:

Use signup flows to gather essential information that will meaningfully personalize the user's first experience. They're your chance to learn about users' goals and preferences right when motivation is highest.

Signup Flow Best Practices:

1. **Keep it Essential:** Ask only what you need to deliver immediate value. Each question should personalize the user's experience. My rule of thumb is to keep it at three questions or fewer.

2. **Show Value and Progress:** Explain how each piece of information will benefit users ("We'll show you relevant templates") and include clear progress indicators ("Step 2 of 4").

3. **Maintain User Control**: Always provide ways to skip steps or go back. Users should feel guided, not trapped, and end with a clear promise of value ("Your workspace is ready!").

2. Empty States: Guide First Actions

Empty states transform blank screens into action-driving moments. Instead of showing "No data yet," they guide users toward their first success.

Example: Loom's empty dashboard shows example videos that both preview the filled state and teach product usage, making the blank canvas less intimidating and more instructional.

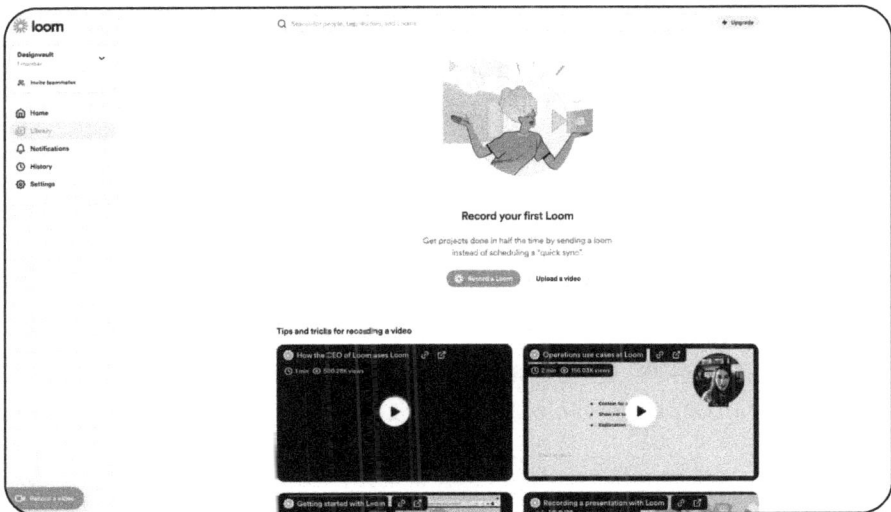

When to Use Empty States:

Empty states are crucial first impression moments—when users encounter a blank dashboard, list, or workspace. These seemingly empty moments are opportunities to guide users toward their first meaningful action.

Empty State Best Practices:

1. **Guide with Clarity and Purpose**: Show one clear next step with an obvious benefit. Instead of "Create project," try "Create your first project to keep your team aligned and on track."

2. **Show the Potential**: Use engaging visuals or examples to help users envision what's possible. Templates and sample data can transform an intimidating blank canvas into an inspiring starting point.

3. **Stay Positive and Focused**: Avoid negative language like "No data yet." Instead, frame empty states as opportunities: "Ready to start your first project?" Keep options focused—too many choices can overwhelm new users.

4. **Personalize the Experience**: Use data collected during signup to pre-populate relevant content. Asana does this brilliantly by showing sample tasks based on role—marketing teams see campaign tasks, while dev teams get sprint planning examples. This eliminates the intimidating blank slate while showing users how the product serves their specific needs.

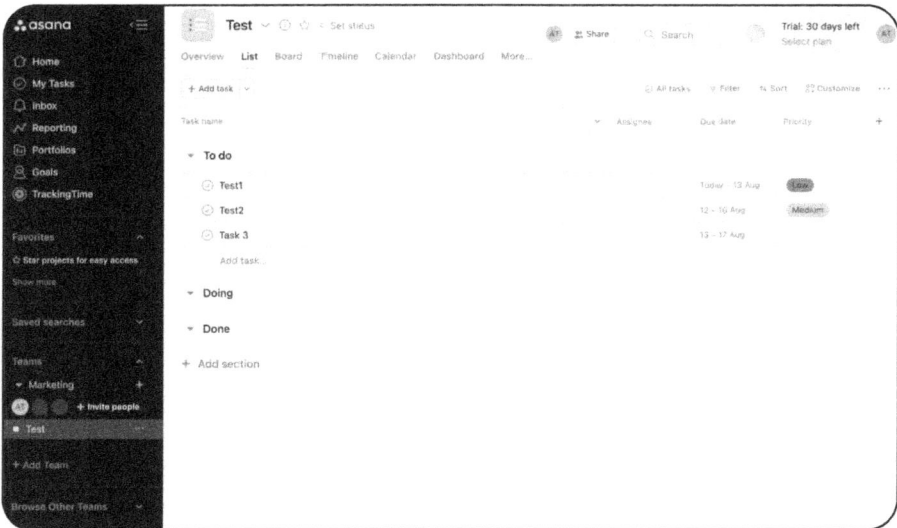

3. Success States: Celebrate and Direct

Success states turn achievements into momentum by celebrating wins while suggesting the next steps. They combine positive reinforcement with clear direction.

> **Example:** Mailchimp pairs celebration with Freddy (their mascot), giving a high five while confirming, "Your mail is in the send queue." It also suggests a next step for the user, "Track your mailing's progress in reports." This keeps users moving forward in their journey.

High Fives!

Your mail is in the send queue and will go out shortly.

Track your mailing's progress in reports

When to Use Success States:

Use success states to celebrate meaningful milestones from your Onboarding Journey Map (Chapter 12). They're perfect moments to reinforce progress while guiding users toward their next achievement.

Success State Best Practices:

1. **Guide Next Steps:** Every celebration should point to a clear next action. When users complete one milestone, immediately show them the path to their next win.

2. **Show the Impact:** Help users understand the value of their achievement, whether it's time saved, results achieved, or progress toward larger goals. When possible, make it shareable to build internal momentum.

3. **Extend the Celebration:** Don't limit celebrations to in-product moments. Follow up with celebratory

emails—they see up to 70 percent open rates because users already feel accomplished. This multi-channel approach reinforces success and maintains momentum.

4. Templates: Accelerate Time-to-Value

Templates help users avoid the anxiety of a blank canvas and jump straight into meaningful work. They provide proven starting points while teaching best practices through examples.

Example: Based on signup responses, Miro presents users with relevant templates for their roles. Product managers see roadmapping, while marketing teams get campaign planning templates. Each comes pre-populated with realistic examples, showing users exactly how to structure their work.

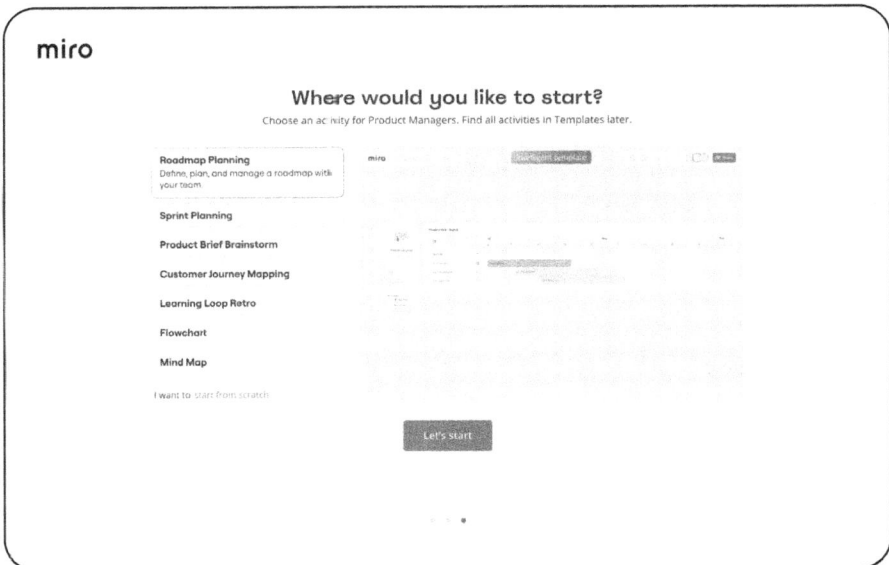

When to Use Templates:

Templates shine wherever you ask users to build something from scratch—whether it's a dashboard, workflow, or project plan. They're particularly valuable when users need to follow specific formats or best practices but might not know where to start.

Template Best Practices:

1. **Start Industry-Specific**: Create templates tailored to different user roles and industries. Marketing teams should see campaign templates, while engineering teams should get sprint planning examples.

2. **Show Real Examples**: Don't just provide empty frameworks. Include realistic sample data that demonstrates how to use the template effectively. This helps users understand both the structure and the context.

3. **Keep It Actionable**: Make templates easy to customize. Users should be able to quickly replace sample content with their own data while maintaining the proven structure.

PILLAR 2: EDUCATIONAL CONTENT

Tell me and I forget, teach me and I may remember, involve me and I learn.

Benjamin Franklin

Software users need more than just basic instructions to become confident experts. While in-product guides provide essential guidance, users need a comprehensive learning environment to master your product. Through various educational resources and content formats, users can build deep product knowledge and develop lasting expertise.

In this section, we'll explore four types of educational content that help users overcome barriers to success:

1. **Onboarding Emails**: Behavior-triggered communication

2. **Knowledge Base**: Self-service documentation

3. **Learning Paths**: Structured education programs

4. **Interactive Demos**: Hands-on practice environments

1. Onboarding Emails: Guide Based on Behavior

Most companies send emails based on time—days one, three, and seven. But best-in-class B2B companies trigger emails based on user actions (or meaningful inactions). This behavior-based approach ensures users receive the right guidance at the right moment.

> **Example:** Loom's first recording success email celebrates the win ("Congratulations, you did it!") and immediately suggests the next step ("Share your recording!"). They follow up with quick use cases like replacing status meetings and giving visual feedback, helping users envision how to make video messaging a daily habit.

☀ loom

Congratulations, you did it! You recorded your first video message with Loom.

You're one step away from greatness: Share your recording! Send your video's URL to spark a conversation with a colleague, client, or friend.

Share your video

Not sure what else to record? Here are some quick ideas to help you be a Loom pro in no time:

- Ditch your next Zoom and send your status update in a loom.
- Send a greeting or congratulate a teammate – they'll appreciate it!
- Send feedback with a loom – it's so much easier than typing it out.

Happy Recording!
The Loom Team

Loom, Inc.
85 2nd Street, Floor 1
San Francisco, CA 94105

When to Use Onboarding Emails:

Send targeted emails when users show interest in features but haven't taken action, miss key milestones, or show signs of disengagement. Your email strategy should respond to user behavior, not just follow a preset timeline.

Onboarding Email Best Practices:

1. **Create Dynamic Journeys:** Replace static drip sequences with behavior-triggered email paths. For example, Nylas increased its free-to-paid conversion by 80 percent by implementing dynamic journeys using Inflection.io. Their system fast-tracks engaged users to advanced content while providing extra guidance to those who need it so users receive the right content at the right moment in their journey.

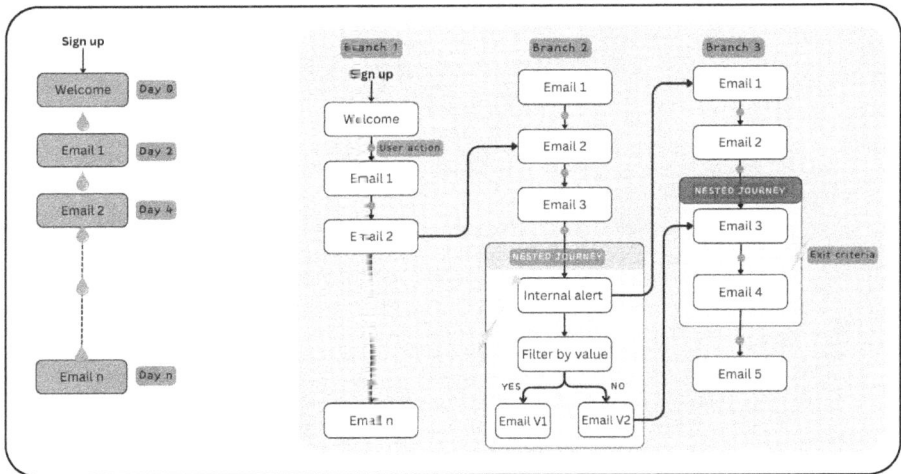

2. **Welcome with Authority:** Make welcome emails personal and confident. Like OptinMonster's co-founder Thomas Griffin, combine social proof ("Over 1,000,000 websites use OptinMonster") with clear next steps and direct support access.

optinm🐹nster

Welcome to OptinMonster!

Thomas Griffin, President and Co-Founder, OptinMonster

You just made a great decision, Ramli!

Hi, Iâ€™m Thomas Griffin, President and Co-Founder of OptinMonster.

Over 1,000,000 websites use OptinMonster to grow their lists, leads and sales, and now youâ€™re one of us, too!

To help you quickly get started, we have created this underline short video that walks you through the four simple steps to publishing your first OptinMonster campaign.

WATCH NOW!

Most people are able to launch their first campaign within a day of registering for OptinMonster. That video will help make sure youâ€™re one of them.

Of course, if youâ€™d rather read than watch a video, our Guide to Creating Your First Campaign **is for you.** It includes an easy to follow, step-by-step tutorial for launching your first campaign, including screenshots.

Iâ€™m going to send you four other emails this week to help equip and inspire you to take massive action using OptinMonster (including a few gifts along the way). Iâ€™d hate for you to miss out on those gifts, thoughâ€¦Could you do me a favor real quick and add my email address to your safe senders list? That will make sure nothing slips through the cracks.

If you run into any challenges as youâ€™re getting started, please donâ€™t hesitate to reply to this email. Iâ€™ll make sure my team takes good care of you.

Talk again soon,

Thomas Griffin

Co-Founder and President of OptinMonster

P.S. Keep an eye out later today for the first gift in your email inbox. I want to make sure you get started on the right foot so I had Siera from our Customer Success team put together something special for you...

Thatâ€™s it!

Now stay tuned for that next emailâ€¦

3. **Celebrate and Guide**: As shown in the Loom example above, pair celebration with clear next steps. Each milestone becomes an opportunity to suggest the next valuable action.

4. **Re-engage with Value**: When users go quiet, offer multiple paths back to value. Notion exemplifies this by providing educational resources, templates, and collaboration options without using guilt or pressure tactics.

From: **Notion Team** <team@mail.notion.so>
Date: Wed, Nov 16, 2022 at 10:31â€¯AM
Subject: A few last thoughts for you ðŸ'

N

Hi there,

Last email from us for a while. We just want to make sure you're all set to use Notion the way you hoped when you signed up. In that spirit, some reminders:

- Want to use Notion with your team? We have a Team Plan that lets you all share a workspace - learn more here
- We're always adding new videos on YouTube - right now, focused on walkthroughs of how to build your own tools and setups in Notion
- If you're switching from another app, we have simple import guides for Evernote, Trello, Google Docs, and many more
- Questions or feedback? Send us an email anytime at team@makenotion.com

Our team is always trying to make the product better. Thanks for being here with us - it means a lot.

Go to Notion

The Notion team

P.S. Our Guides & FAQs are always here for you.

N Notion

Notion.so, the all-in-one workspace for
your notes, tasks, wikis, and databases.

548 Market St #74567, San Francisco, CA 94104

Unsubscribe

FREE RESOURCE

Download proven behavior-based email templates and timing strategies at eurekabonus.com. These templates include trial expiration reminders, missed milestone follow-ups, and declining usage interventions.

2. Knowledge Base: Enable Self-Service Learning

Your knowledge base serves as your product's 24/7 self-service library. The key isn't just having comprehensive documentation—it's making information easy to find and follow.

Example: Slack's help center exemplifies this with a clear structure: prominent search, common troubleshooting topics, and well-organized categories. Their "Getting Started" section guides new users through basics, while advanced documentation helps power users customize their workspace.

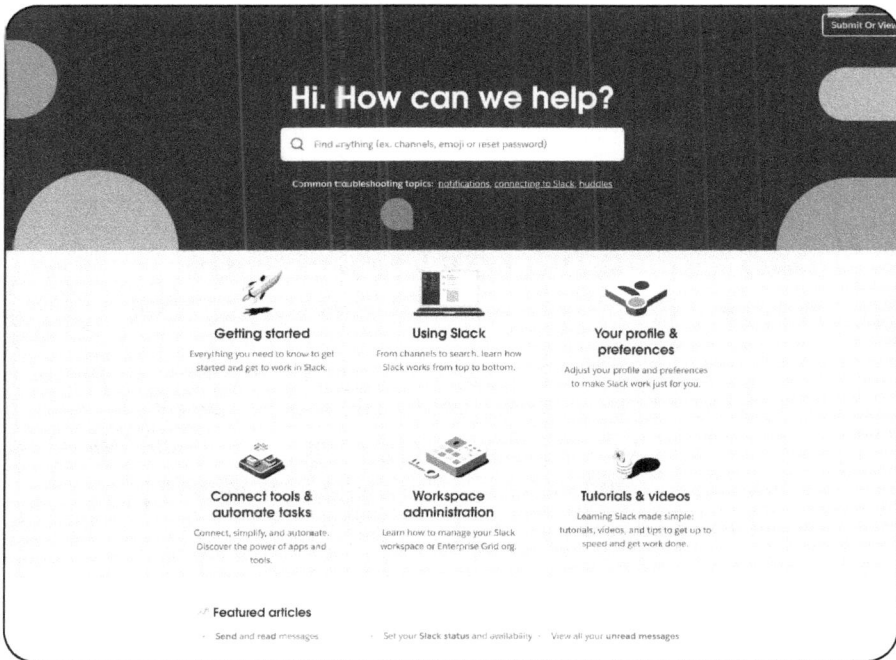

When to Use Knowledge Base:

Make your knowledge base the go-to resource for product documentation, troubleshooting guides, and best practices. It should answer both common questions from new users and complex queries from power users.

Knowledge Base Best Practices:

1. **Structure for Discovery**: Organize content clearly with intuitive categories and powerful search. Users should find answers in seconds, not minutes.

2. **Show, Don't Just Tell**: Include visual aids like GIFs, screenshots, and video walk-throughs. Complex features often need demonstration, not just description.

3. **Keep Content Fresh**: Regularly update documentation
    as your product evolves. Outdated content frustrates
    users and increases support tickets.

PRO TIP: PRIORITIZE KNOWLEDGE BASE CONTENT BASED ON FRICTION

*Use your Friction Map from Chapter 13 to prioritize
which knowledge-base articles to create first. Focus on
addressing common friction points you've identified in
your user journey.*

3. Learning Paths: Build Lasting Expertise

While knowledge bases offer quick answers, learning paths provide
structured education programs that build lasting expertise. Think of
it like the difference between looking up a recipe versus attending
cooking school.

Example: HubSpot Academy combines product mastery with
professional growth. Its role-based certifications (Marketing,
Sales, Service) teach more than HubSpot features—they teach
industry best practices that make users more valuable to their
organizations.

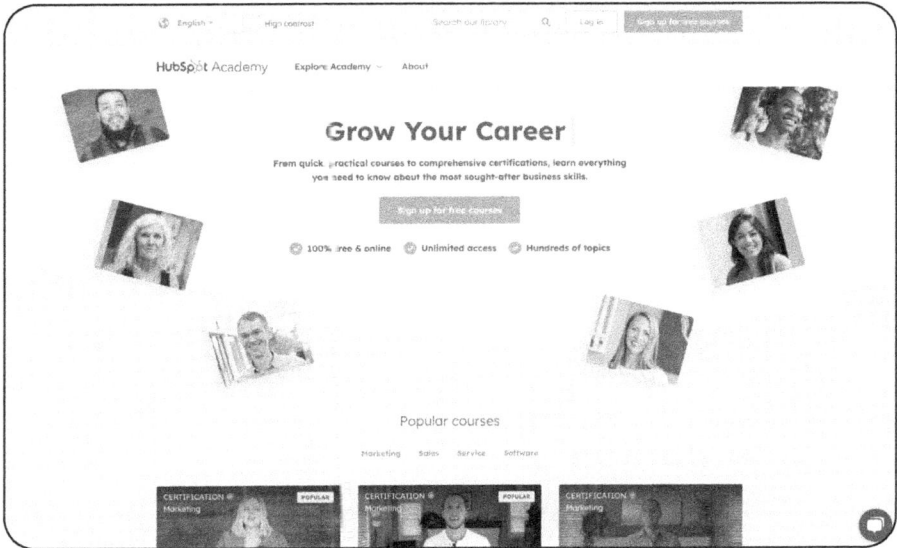

When to Use Learning Paths:

Create learning paths when users need a comprehensive understanding of complex workflows or mastery requires structured progression through concepts. They're particularly valuable for enterprise products requiring product and industry expertise.

Learning Path Best Practices:

1. **Design Role-Based Tracks**: Create focused paths for different roles and skill levels. Marketing teams need different training than sales teams.

2. **Balance Theory and Practice**: Combine concept explanations with hands-on exercises. Users should learn by doing, not just watching.

3. **Validate Progress**: Offer certificates or assessments that users can share. This builds confidence and helps them demonstrate value to their organizations.

PRO TIP: REPURPOSE ACADEMY CONTENT

Follow HubSpot's lead by breaking down academy videos into shorter snippets for in-product guides. These "on-the-go" tutorials provide quick help while encouraging users to visit the academy for deeper learning. This approach creates a seamless connection between immediate guidance and comprehensive education.

Further Resources:

Customer education has evolved into its own field with dedicated experts and resources. For a comprehensive look at building education programs, I recommend *Customer Education* by Adam Avramescu, which provides an excellent framework for building scalable education programs. I also recommend checking out SaaS Academy Advisors (saasacademyadvisors.com), run by Chris LoDolce and Lindsay Thibeault (who previously worked at HubSpot Academy). These resources go deeper into education strategy than we can cover here, but they're invaluable if you're building a comprehensive education program.

4. Interactive Demos: Enable Learning by Doing

Interactive demos let users experience your product's value hands-on before committing. Unlike passive videos, they create muscle memory through guided practice.

Example: Remote's in-product demo library lets users choose interactive tours based on their needs.

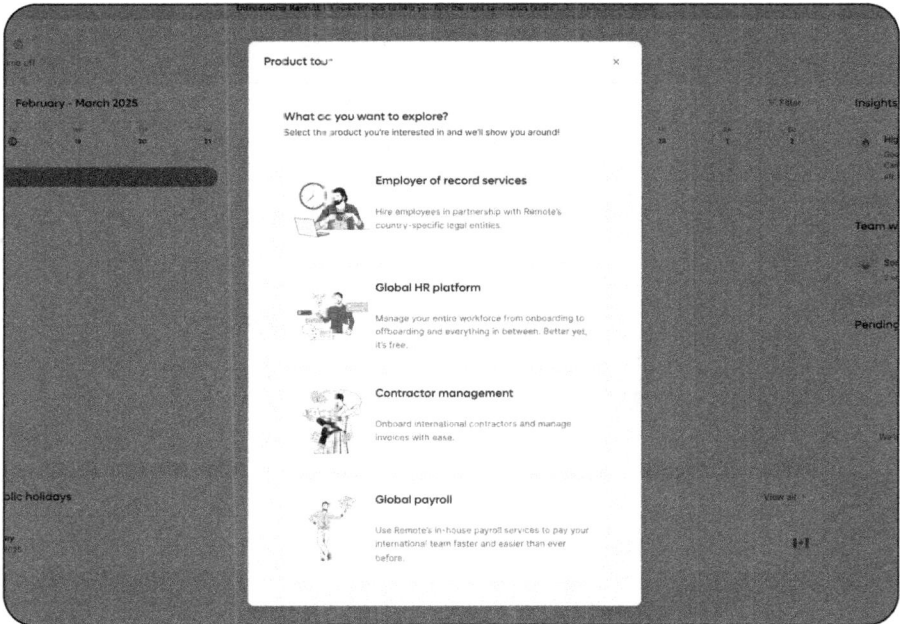

For example, they showcase how new employee onboarding looks from start to finish using an interactive demo. By walking users through a complete hire with sample data, they reduce the emotional friction and anxiety around managing international employees—users can see exactly how Remote handles compliance, payroll, and benefits before adding their first team member.

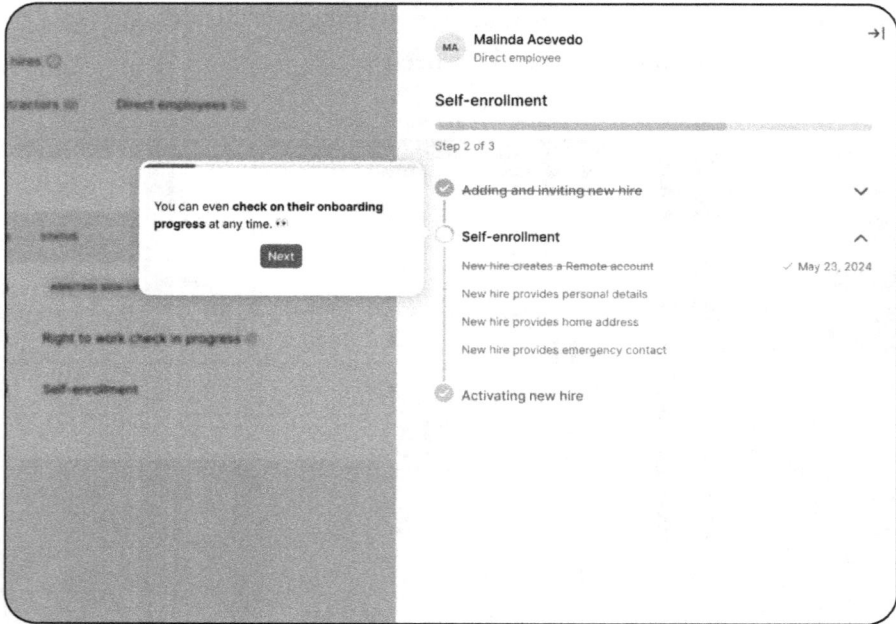

When to Use Interactive Demos:

Use interactive demos to showcase complex workflows or when users need hands-on practice in a safe environment. They're particularly effective when users want to test features before implementing them.

Interactive Demo Best Practices:

1. **Focus on Value**: Show users achieving meaningful outcomes, not just clicking through features. Use realistic, industry-specific data that resonates with your audience.

2. **Keep It Focused**: Guide users through one clear workflow at a time. Don't try to showcase every feature; instead, focus on the most impactful actions.

PRO TIP: MATCH DEMOS TO AHA MOMENTS

"Match your interactive demos to key aha moments or moments of delight in the product. With an interactive demo, the goal is to show users an ideal end state that gets them excited about your product and going through onboarding."

Natalie Marcotullio, Head of Growth at Navattic

PILLAR 3: HUMAN INTERACTION

Technology is a useful servant but a dangerous master. The more technology we have, the more we need to be human.

Christian Lous Lange, Historian and Nobel Peace Prize Winner

During my first week of college, I was completely lost trying to understand calculus. The textbook explanations and online videos weren't clicking. Then, I attended my professor's office hours. In fifteen minutes of human interaction, she helped me grasp concepts I'd been struggling with for days. She didn't just explain the formulas—she understood where I was getting stuck and adapted her explanation to my needs.

The same principle applies to B2B onboarding. While automated guides provide foundations, certain situations need the adaptability and empathy only human support can provide. In this section, we'll explore one-to-many approaches (like webinars and community programs) and one-to-one support (high-touch onboarding).

Scaling Human Support Effectively

Before diving into specific approaches, it's crucial to determine which type of human interaction best serves your users. Two key factors influence this decision:

1. **Product Engagement**: How actively users complete key actions from your Journey Map

2. **Annual Contract Value (ACV)**: The revenue potential that justifies personalized attention
   Looking at these factors in a matrix helps determine the right approach:

   ○ **High Engagement + High ACV**: Ideal for one-on-one, high-touch onboarding. These customers deserve dedicated implementation support to accelerate their success.

   ○ **High Engagement + Low ACV**: Perfect for one-to-many approaches like webinars and community programs. These engaged users often become product champions and community leaders.

   ○ **Low Engagement + Low ACV**: Focus on improving automated solutions—better in-product guides and educational content.

   ○ **Low Engagement + High ACV**: Prioritize immediate one-to-one intervention to understand and address engagement barriers.

One-to-many
Onboarding

One-to-One
High-Touch Onboarding

High

Product
Engagement

Low

Automate up

Nurture and automate
to increase
product engagement

Red Flag

Prioritize immediate
one-to-one
human intervention

Low   Annual Contract   High
Value

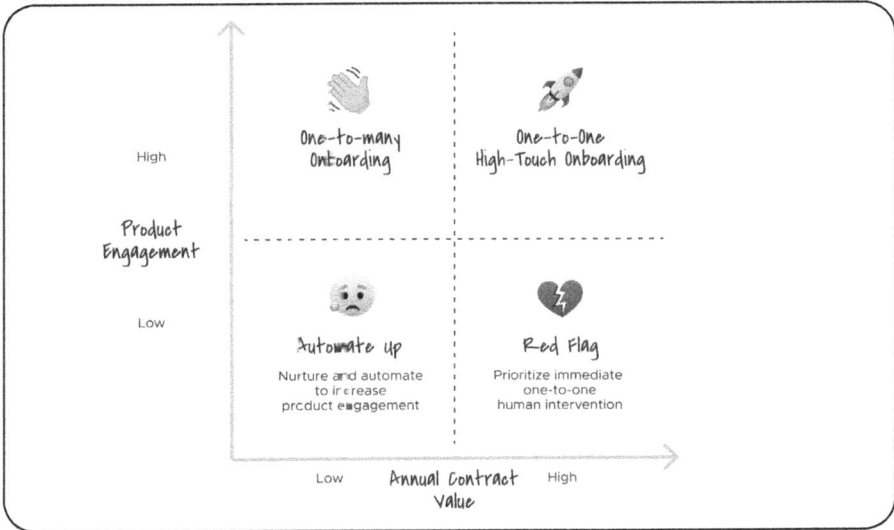

Let's examine each human interaction approach, starting with scalable group training through webinars.

1. Onboarding Webinars: Scale Human Connection

As your user base grows, individual training becomes unsustainable. Onboarding webinars combine the efficiency of one-to-many training with the warmth of human interaction, creating opportunities for real-time learning and community building.

Example: Asana Academy's live training program demonstrates the power of structured webinars that grow with users' needs:

Their program combines skill levels with specific use cases:

- Beginner sessions teach core task management.
- Intermediate sessions help teams get organized together.
- Advanced sessions tackle cross-functional scenarios.

Each sixty-minute session includes demos, Q&A, and recordings, available in multiple languages for global teams.

When to Use Onboarding Webinars:

Use webinars to train multiple users on common workflows or when concepts benefit from live demonstrations and Q&A. They're particularly effective for teaching best practices that might not be obvious from the product interface alone.

Onboarding Webinar Best Practices:

1. **Structure for Impact:** Divide the sixty-minute session into clear segments: five minutes for the welcome, fifteen

minutes for core concepts, twenty minutes for a live demonstration, fifteen minutes for Q&A, and five minutes for the next steps.

2. **Keep It Interactive**: Start with icebreaker polls and maintain engagement through chat discussions. Use real examples instead of slides, and have a co-host manage audience interaction.

3. **Follow Up Thoughtfully**: Within twenty-four hours, send recordings and resources, along with clear next steps and a feedback survey. Invite participants to future sessions while their interest is high.

Further Resources:

For deeper guidance on creating engaging virtual training sessions, I recommend exploring Roger Courville's *The Virtual Presenter's Handbook* and Rob Fitzpatrick's *The Workshop Survival Guide*. These authors offer practical frameworks for designing and delivering interactive sessions that keep participants engaged and learning effectively—going well beyond what we can cover in this chapter.

2. Community Programs: Enable Peer Learning

While webinars provide structured learning, community programs create ongoing peer-to-peer support and learning opportunities. The key isn't the platform—it's creating a space where users want to help each other succeed.

Example: Circle uses its own community platform to onboard new customers through its "Share & Learn" channel. Instead of just providing documentation, they encourage users to share their experiences, milestones, and tips.

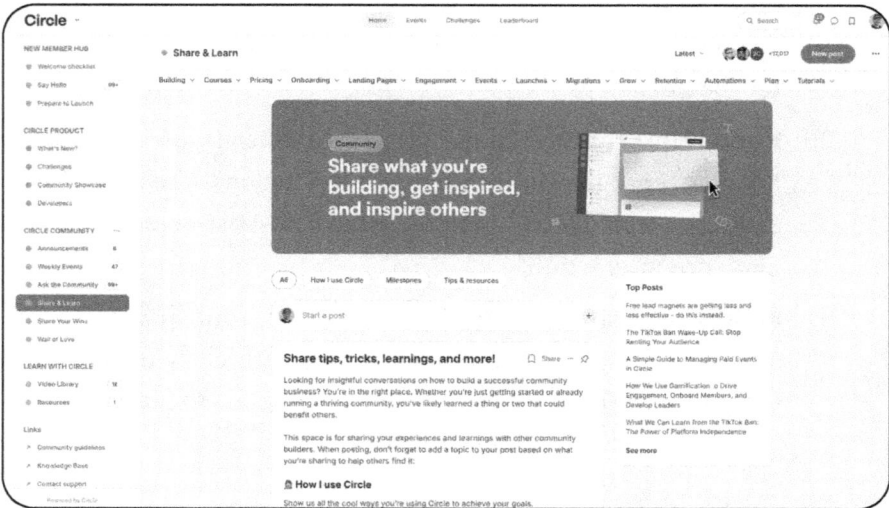

The "Video Library" channel offers expert content through masterclasses, customer showcases, and workshops, which are available both on-demand and live.

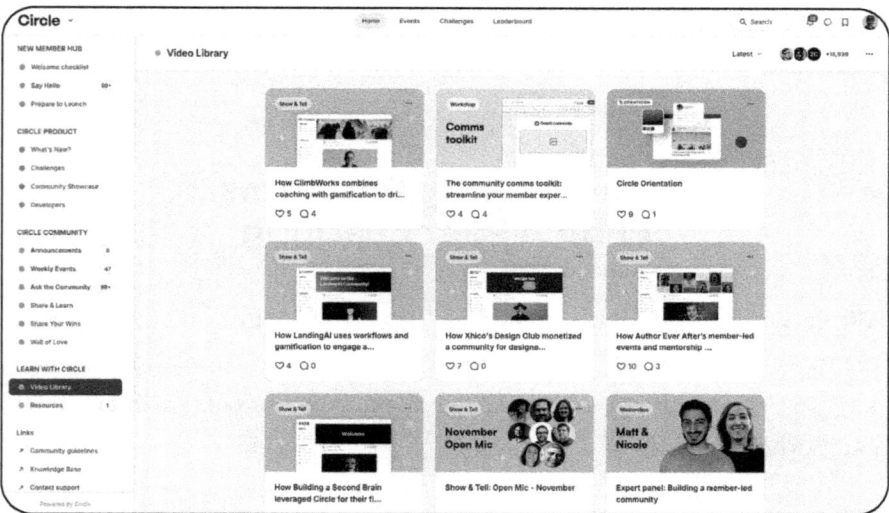

When to Use Community Programs:

Consider launching a community when:

- Your product has multiple paths to success.
- Users can learn valuable lessons from each other.
- You have resources for consistent community management.
- You're ready for long-term community building.

Community Best Practices:

1. **Focus on Value Exchange**: Create spaces where users can both contribute and benefit. Encourage sharing of real experiences and practical tips rather than just asking questions.

2. **Build Active Engagement**: Welcome new members personally and celebrate community wins. Have clear guidelines that encourage meaningful interactions while maintaining focus.

3. **Stay Consistently Present**: Community-building requires daily attention. Only start if you're ready to invest in nurturing genuine connections and supporting members' growth.

Further Resources:

Building and scaling communities is a complex topic that I can't do justice to in this book. For deeper guidance, I recommend exploring works like Charles Vogl's *The Art of Community*, Peter Block's *Community: The Structure of Belonging*, and David Spinks's *The Business of Belonging*. These authors offer comprehensive frameworks for creating genuine community engagement beyond what we can cover here.

3. High-Touch Onboarding: Guide Complex Implementations

While webinars and communities help scale support, enterprise accounts and complex implementations need dedicated attention. High-touch onboarding provides personalized guidance for successful adoption across large organizations.

Example: Appcues runs a structured 12-week program focused on customer independence:

1. Weeks one through four: Technical setup and basic training

2. Weeks 5 – 12: Best practices and go-live

3. Week 13: Transition to Success team

STAGE	TIME PERIOD		EVENT	DESCRIPTION
Technical Implementation and Basic Training	Month 1	Week 1	Onboarding Kickoff	Share the onboarding program overview for the customer and their team. Confirm the customer's use case(s). Set expectations for the next 45 days. Verify the benchmarks. Schedule the training and implementation sessions.
		Week 2	Pre-Installation Call	Review installation requirements, integrations, and kickoff discussions.
		Week 3	Basic Platform Training	Training on basic setup to get the customer to started building flows.
		Week 4	Post-Installation Call	Post-installation review call to confirm successful installation. Audit the data and discuss integrations
Best Practices and Go-Live Support	Month 2-3	Week 5	Best Practices Training	Training on use case-focused features
		Week 7	Use Case and ROI Review	Confirm ROI metrics with the customer including benchmarks and goals.
		Week 9	Go-Live Support 1	Open agenda call to answer customer questions, review content, and provide best practices.
		Week 11	Go-Live Support 2	Open agenda call to answer customer questions, review content, and provide best practices.
CSM Introduction	Month 4	Week 12	CSM Handoff	CSM intorduction, where the CSM discusses the success plan and transition handoff.

When to Use High-Touch Onboarding:

Reserve high-touch onboarding for situations that demand dedicated attention, particularly in these situations:

1. **High-Value Accounts:** Enterprise customers require careful coordination across departments and customized support.

2. **Complex Implementation Needs:** When implementations involve multiple stakeholders, technical requirements, and change management challenges, hands-on guidance becomes crucial. From coordinating across teams to handling custom integrations, these situations need human support for successful outcomes.

3. **Risk Signals:** User behavior can signal the need for intervention: extended trial inactivity, technical roadblocks, or repeated failed attempts at key actions. Quick human response in these moments can prevent abandonment and maintain implementation momentum.

I suggest that you and your team define when to engage with human support based on clear criteria, such as:

- Company size (e.g., 500+ employees or $50M+ revenue)
- Technical complexity (e.g., SSO implementation, custom API needs)
- Implementation scope (e.g., 100+ user rollout, multi-department deployment)

High-Touch Best Practices:

1. **Start with Clear Handoff:** The moment after a sale closes is crucial. Document exact customer goals, timeline commitments, and technical requirements. Have both Sales and Customer Success teams present to ensure alignment.

2. **Structure the Journey:** Break implementation into clear phases with defined milestones. Celebrate quick wins (like successful technical setup) while building toward larger goals.

3. **Enable Independence:** Focus on teaching rather than doing. By program's end, customers should be able to manage their implementation independently and have clear ROI metrics.

PRO TIP: CREATE A SALES-TO-ONBOARDING HANDOFF AGREEMENT

Have Sales and Customer Success sign off on a formal document covering goals, timelines, requirements, stakeholders, and expectations. Download templates at eurekabonus.com.

Further Resources:

Implementing human-touch onboarding deserves deeper study. I recommend exploring Donna Weber's *Onboarding Matters* and *The Customer Success Professional's Handbook* by Ashvin

Vaidyanathan and Ruben Rabago for comprehensive frameworks on building effective onboarding programs and customer success teams. These books provide detailed guidance on scaling human interactions effectively while maintaining high-touch experiences.

WHERE TO GET STARTED

Rather than implementing all onboarding tactics at once, use your Friction Map from Chapter 13 to prioritize your efforts. Here's how to begin:

1. Address Top Issues First

Start by reviewing your Friction Map and support tickets to identify where users most commonly get stuck. These pain points will guide which tools to implement first.

2. Match Solutions to Friction Types

Different types of friction often require a combination of tools. Functional friction—like complex setup processes or feature workflows—can be addressed through multiple approaches. While in-product guides provide immediate help, learning paths offer deeper understanding, and webinars allow for live demonstrations and Q&A.

Social friction, like driving team-wide adoption or managing organizational change, needs a multi-faceted approach. While human guidance helps navigate organizational politics, educational content plays a crucial role, too. Implementation playbooks give teams clear rollout strategies, customer case studies build internal confidence, and learning paths help create internal champions who can drive adoption.

Community programs can also help by connecting teams with peers who've successfully managed similar transitions.

Emotional friction, such as anxiety about managing international payroll or uncertainty about compliance, requires both reassurance and proof. Human interaction provides the personal guidance and industry expertise that builds confidence. But combine this with educational resources like detailed guides, success stories, and interactive demos that let users practice in safe environments. Webinars can also help by showing real examples and letting users ask questions about their specific concerns.

3. Start Small and Iterate

Begin with one tool in your highest-impact area. For example, if users struggle with initial setup, start with targeted tooltips rather than building a complete academy. Measure the impact, gather feedback, and refine your approach before expanding to other areas.

Remember: The goal isn't to implement every tool—it's to provide the right combination of support that helps users overcome their specific barriers to success. In the next chapter, we'll explore how to measure the effectiveness of these implementations and create feedback loops for continuous improvement.

In the next chapter, we'll put these tools into action through the Friction-to-Action Workshop. You'll learn how to transform your friction points into concrete experiments, evaluate solutions based on impact and effort, and create clear action plans that leverage the right combination of tools for your specific challenges.

STEP 5

Apply, Analyze, and Repeat

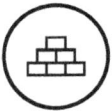

Establish
a Team

Understand
Success

Reverse
Journey Map

Keep Users
Engaged

Apply Analyze
and Repeat

16

The Friction-to-Action Onboarding Workshop

~~~

Obstacles don't have to stop you. If you run into a wall, don't turn around and give up. Figure out how to climb it, go through it, or work around it.

Michael Jordan

~~~

For my thirty-ninth birthday, my wife took me to Eleven Madison Park, the three-Michelin-starred restaurant in New York. I had just finished reading *Unreasonable Hospitality* by Will Guidara, EMP's former General Manager, who transformed fine dining from a transaction into an unforgettable experience.

Guidara writes about how he would gather his team to identify every moment that felt too transactional in the dining experience. They'd brainstorm dozens of ideas, then carefully prioritize which improvements to implement first. Some were quick wins, like moving the maître d' from behind the podium to greet guests at the door—a simple change that immediately made arrivals feel more welcoming. Others required more significant investment, like implementing a ticketless coat check system that eliminated the awkward fumbling for claim tickets at the end of meals.

Like Guidara's approach to hospitality, improving product onboarding requires methodical prioritization. You can't fix everything at once.

Instead, you need to identify friction points and carefully choose which improvements will have the biggest impact.

We've now reached the final phase of the EUREKA Framework: Apply, Analyze, and Repeat. After mapping our friction points and exploring the three pillars of successful onboarding in previous steps, it's time to put these insights into action and create a continuous improvement cycle. The Friction-to-Action Workshop will help you generate creative solutions, evaluate them based on impact and effort, and create clear action plans.

THE FRICTION-TO-ACTION ONBOARDING WORKSHOP

This is where everything comes together. The Friction-to-Action Workshop is your "putting it all together" moment—where you'll transform the friction points you've identified and the solutions you've explored into a concrete action plan. Through this rapid-fire workshop, you'll generate and prioritize solutions across the three pillars of onboarding, creating a clear roadmap to help users reach their activation goals faster. It's also where you'll establish the foundation for continuous improvement, setting up ways to measure success and iterate on your solutions.

Before You Begin

Before running this workshop, have your team:

- Review your completed Friction Map from Chapter 13
- Familiarize themselves with Chapter 15, where I covered the three pillars of successful onboarding—in-product guides, educational content, and human touchpoints

- Consider common solution patterns for each friction type:

  1. Functional friction: product tours, tooltips, documentation
  2. Social friction: implementation playbooks, change management guides
  3. Emotional friction: success stories, confidence-building quick wins

The Setup

Time needed: 30 minutes

Participants: Your cross-functional onboarding team (4–8 people)

Materials needed:

- Your completed Friction Map

- Square sticky notes (three different colors)

- Sharpies or markers

- Voting dots (10 per person)

- Timer

- Wall space or digital whiteboard

You can download templates for FigJam and Miro, along with facilitation guides and example solutions, at *eurekabonus.com*. These digital templates are particularly helpful for remote teams or distributed workspaces.

The Activity

Step 1: Generate Solutions (10 minutes)

The key to this exercise is working "Together Alone"—while everyone shares the same goal of improving onboarding, each person generates ideas independently without discussion or sharing. This approach prevents groupthink and ensures all voices are heard, not just the loudest ones.

1. Frame your challenge as a "How Might We" (HMW) question based on your top friction point from Chapter 13. For example:

   - "How might we help users complete their first data import?"
   - "How might we increase team-wide adoption?"
   - "How might we build user confidence during setup?"

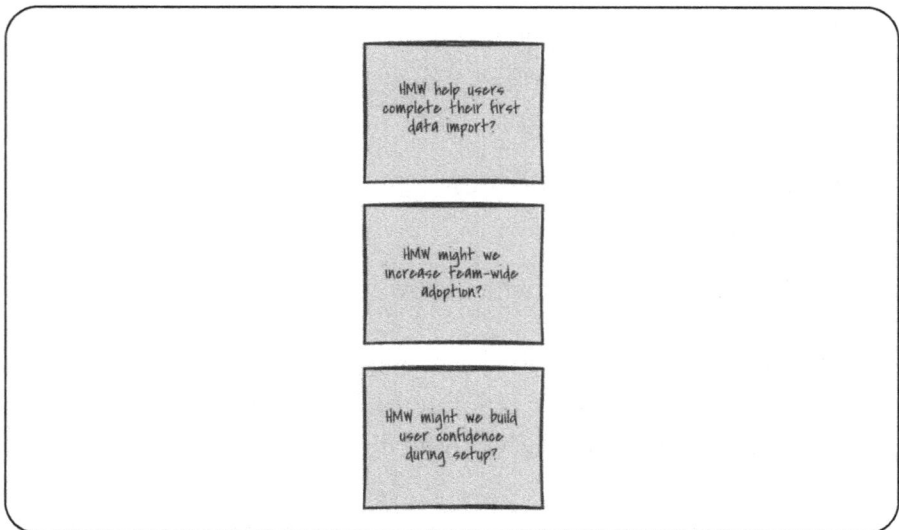

2. Give each participant a stack of sticky notes and a marker. Consider playing background music to make the silence more comfortable.

3. Explain the ideation rules:
    - Write one idea per sticky note.
    - Aim for at least 15 ideas (quantity over quality!).
    - Consider solutions across all three pillars.
    - Write legibly—these will be shared anonymously.
    - No discussion during the ideation step.

4. Set a timer for 10 minutes and begin silent ideation.

PRO TIP: PUSH THROUGH THE IDEA WALL

The purpose of this step is to generate as many potential solutions as possible. Don't worry about feasibility yet—that comes later. When participants hit the "idea wall" and stop writing, encourage them to keep going. To spark creativity, ask:

- *"What would a different industry do?"*
- *"What's the opposite approach?"*
- *"What's the smallest version?"*
- *"What if resources were unlimited?"*

Step 2: Curate and Share (5 minutes)

Every participant should now have a healthy stack of potential solutions. Many ideas won't be feasible—and that's perfectly fine! The goal of this step is to identify the most promising solutions to pursue.

1. Review your ideas silently. Take 2 minutes to read through everything you've generated.

2. Select your top 10 solutions that best address the "How Might We" question. Be ruthless in your selection—only the most promising ideas should make the cut.

3. Discard all remaining ideas—yes, actually throw them away! This will help you maintain focus on your strongest solutions.

4. Post your chosen ideas randomly on the wall or in a digital workspace. Don't group or organize them—the mix helps maintain anonymity.

5. The facilitator should quickly remove any duplicate ideas that emerge.

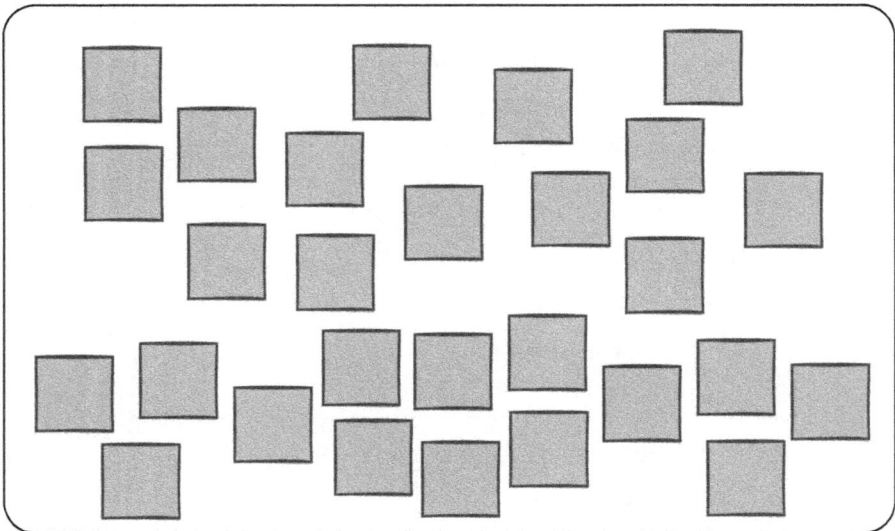

PRO TIP: IMPACT-FIRST SELECTION

When selecting your top 10 ideas, remind participants to focus on potential impact rather than ease of implementation. Ask them to consider:

- *"Which solutions best address our 'How Might We' question?"*
- *"What would make the biggest difference for users?"*
- *"What tackles root causes, not just symptoms?"*

Include at least 2–3 bold, game-changing ideas—even if they seem challenging to implement.

Step 3: Silent Voting (5 minutes)

Rapid dot voting helps identify the most promising solutions while avoiding lengthy discussions. The team's collective wisdom emerges through quick, gut-reaction choices rather than debate.

1. Remove any remaining duplicate ideas from the board. Take a moment to review the original "How Might We" question together, ensuring everyone stays focused on the core challenge.

2. Give each participant 10 voting dots. For digital workspaces like FigJam or Miro, use their built-in voting features.

3. Explain the voting rules:
   - Place dots on ideas you think are most promising.
   - Can put multiple dots on a single idea.

- ○ Can vote on your own ideas.
- ○ If an idea isn't clear, skip it (no asking for clarification).
- ○ Must use all 10 dots within the time limit.
- ○ No discussion during voting.

4. Set a timer for 3 minutes and begin silent voting. Encourage participants to trust their gut reactions rather than overthinking each choice.

5. Use the final minute to count dots per idea quickly. Keep all ideas visible for now—you'll evaluate the top-voted ones in detail during the Impact/Effort matrix in Step 4.

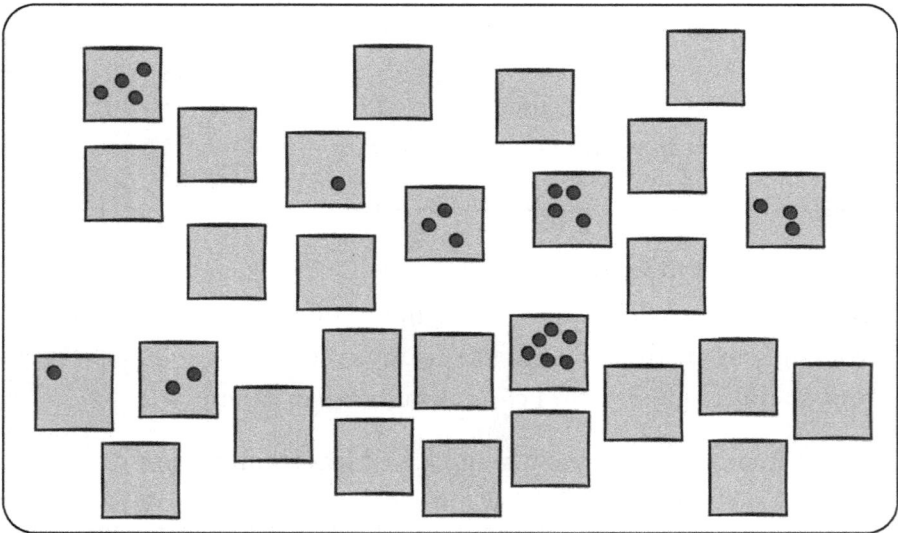

Step 4: Prioritize with Impact/Effort Matrix (15 minutes)

The Impact/Effort Matrix helps transform your voted solutions into an actionable roadmap. This step creates clear priorities and implementation timelines for your onboarding improvements.

1. Organize and arrange the ideas so that the ideas with the most votes are at the top and the ideas with no votes are removed from the board. Take your top 10–12 voted ideas forward and keep the rest for future consideration.

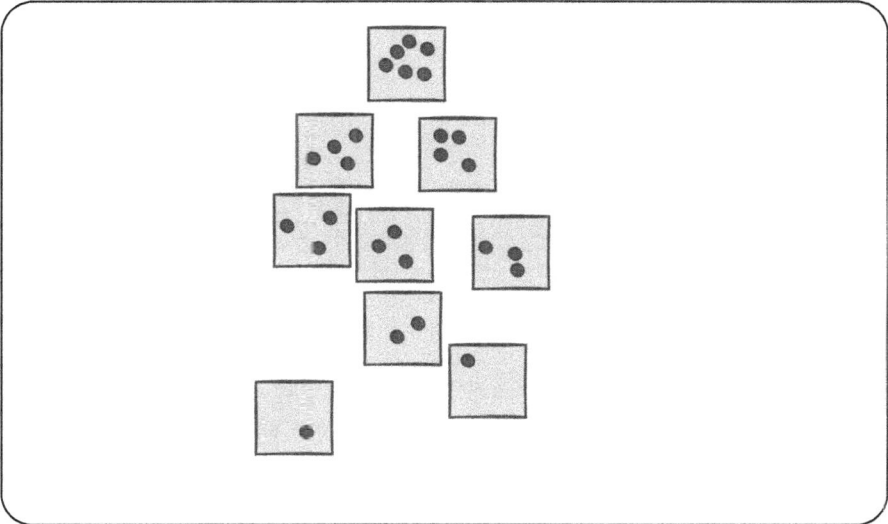

2. Draw a 2x2 grid on your workspace. Label the vertical axis "Impact" (low to high) and the horizontal axis "Effort" (low to high). Mark your quadrants:

    o   Top Left: "Do Now" (High Impact, Low Effort)
    o   Top Right: "Project" (High Impact, High Effort)
    o   Bottom Left: "Quick Wins" (Low Impact, Low Effort)
    o   Bottom Right: "Don't Do" (Low Impact, High Effort)

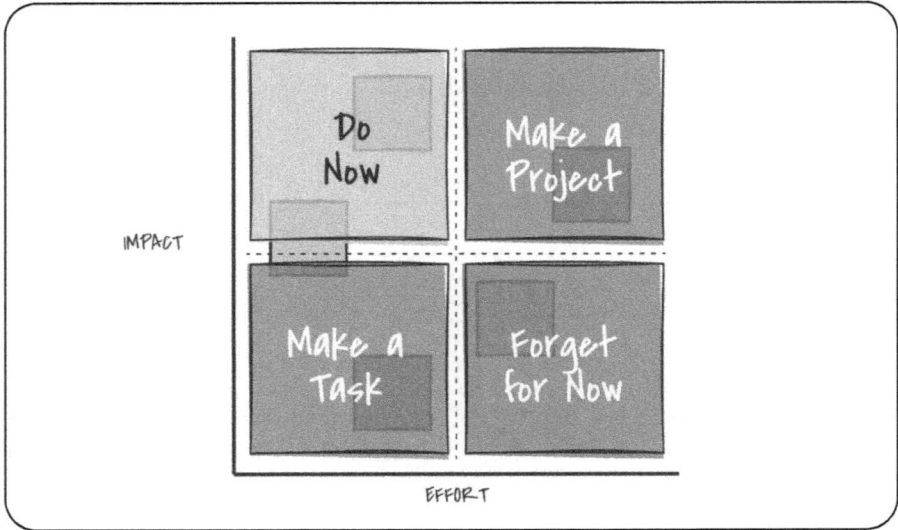

3. Starting with your highest-voted ideas, hold each sticky note in the center of the matrix. Ask the team two questions: "Is this higher or lower impact than average?" followed by "Is this higher or lower effort than average?" Repeat it for the other ideas. Place each solution in its appropriate quadrant based on the team's responses.

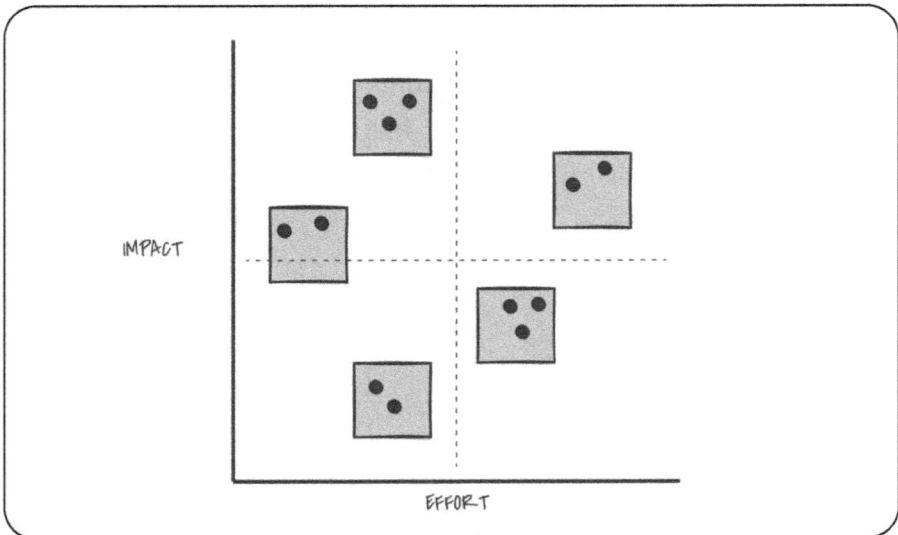

PRO TIP: FOCUS ON ALIGNMENT, NOT PRECISION

Don't get caught up in exact measurements—effort and impact are relative to your context. For instance, creating an academy is high effort for most teams but medium effort if you have existing content to repurpose. Focus on team consensus about where solutions rank relative to each other.

Step 5: Action and Assign (20 minutes)

Transform your prioritized solutions into concrete experiments, starting with your "Do Now" quadrant items. These high-impact, low-effort solutions will generate quick wins for your onboarding.

1. For each "Do Now" solution (focus on the top 2–3), create a clear experiment description. Define what specific change will be made and how it will be implemented, and set a timeframe (aim for two-week experiments). Include specific success criteria with measurable targets. Here's an example action plan:

Solution: "Add progress checklist to onboarding"

-   Experiment: Create 5-item checklist showing key setup steps

-   Implementation: Use Appcues to add checklist to dashboard

- Success Criteria: 50 percent completion rate, 25 percent faster time to value

2. Use the RACI framework to assign clear ownership and accountability. RACI stands for:

    ○ **R**esponsible: Who will do the work
    ○ **A**ccountable: Who owns the results and makes final decisions
    ○ **C**onsulted: Who provides input and expertise
    ○ **I**nformed: Who needs to be kept updated

3. Here's an example for improving a project management tool's onboarding, where the Ultimate Win is "Run first data-driven status meeting":

    Example Solution: "Add a progress checklist to guide users into the first status report."

    Responsible:

    ○ PM: Design checklist
    ○ UX: Create UI
    ○ Writer: In-product copy
    ○ Engineer: Implementation

4. Accountable:

    ○ Head of Growth: Owns initiative and results

5. Consulted:

    ○ Customer Success: User insights
    ○ Sales: Prospect feedback
    ○ Data Team: Success metrics
    ○ Senior Engineer: Technical input

6. Informed:

    ○ VP Product: Monthly metrics
    ○ Marketing & Support: Documentation
    ○ Customer Success: Outreach planning

7.   Document each experiment in your project management tool, including timeline, success metrics, and key stakeholders. Set regular check-in dates to monitor progress.

8.   Document high-impact, high-effort projects in your project management tool to form your longer-term onboarding roadmap. While you focus on implementing quick wins now, these larger initiatives create a vision for your onboarding's future. They can be fleshed out, resourced, and scheduled as your team builds momentum from earlier improvements.

PRO TIP: START SMALL, LEARN FAST

While it's tempting to tackle multiple solutions at once, focus on one experiment at a time. This allows you to measure impact clearly and apply learnings to future implementations. A successful small change often teaches you more than a complex project that's difficult to measure.

You'll now have concrete next steps to start improving your onboarding experience with clear ownership and success metrics.

CONCLUSION

Like Eleven Madison Park's journey to excellence, improving your onboarding experience is an ongoing process of identifying friction points, generating solutions, and carefully implementing changes. The Friction-to-Action Workshop provides a structured approach to this challenge, helping teams move from insight to action.

Remember these key principles as you implement your solutions:

- Start with high-impact, low-effort wins to build momentum.

- Use the three pillars of onboarding to create comprehensive solutions.

- Assign clear ownership using the RACI framework.

- Focus on one or two experiments at a time.

- Document your approach and results.

However, identifying and implementing solutions is only part of the journey. To create lasting improvement, you need to measure the impact of your changes and continuously refine your approach. In the next chapter, we'll explore how to set up effective measurement systems and create feedback loops that drive ongoing optimization of your onboarding experience.

EUREKA ACTION ITEMS: THE FRICTION-TO-ACTION WORKSHOP

Ready to transform your friction points into an actionable onboarding improvement roadmap? Download the workshop templates and facilitation guide from *eurekabonus.com* and follow these steps:

1. **Prepare Your Workshop.**

   - Gather your cross-functional onboarding team.
   - Review your Friction Map from Chapter 13.
   - Familiarize yourself with solutions from Chapters 14–16.
   - Set aside sixty minutes of uninterrupted time.

2. **Generate Solutions.**

   - Frame your top friction point as a "How Might We" question.
   - Run the 10-for-10 exercise to generate ideas.
   - Remember to consider solutions across all three pillars.

3. **Prioritize and Plan.**

   - Create your Impact/Effort matrix.
   - Plot your solutions.

- Identify quick wins to tackle first.
- Document longer-term projects for future planning.

4. Assign Clear Ownership.

- Use the RACI framework for each solution.
- Set clear success metrics.
- Schedule regular check-ins.
- Document everything in your project management tool.

Remember: Start small with one or two high-impact, low-effort solutions. Build momentum with quick wins before tackling larger projects. In the next chapter, we'll explore how to measure the impact of these improvements and create feedback loops for continuous optimization.

17

The Paths to Personalized Onboarding

~~~

Personalization is the key to cutting through the noise and making meaningful connections with customers.

Angela Ahrendts, former SVP of Retail at Apple

~~~

I **visited IKEA to** buy a new computer chair. I knew exactly what I wanted—I'd researched the model online, confirmed it was in stock, and planned to get in and out quickly. But my wife Joanna had other ideas. She wanted to follow IKEA's famous guided path, exploring room displays and discovering new items along the way.

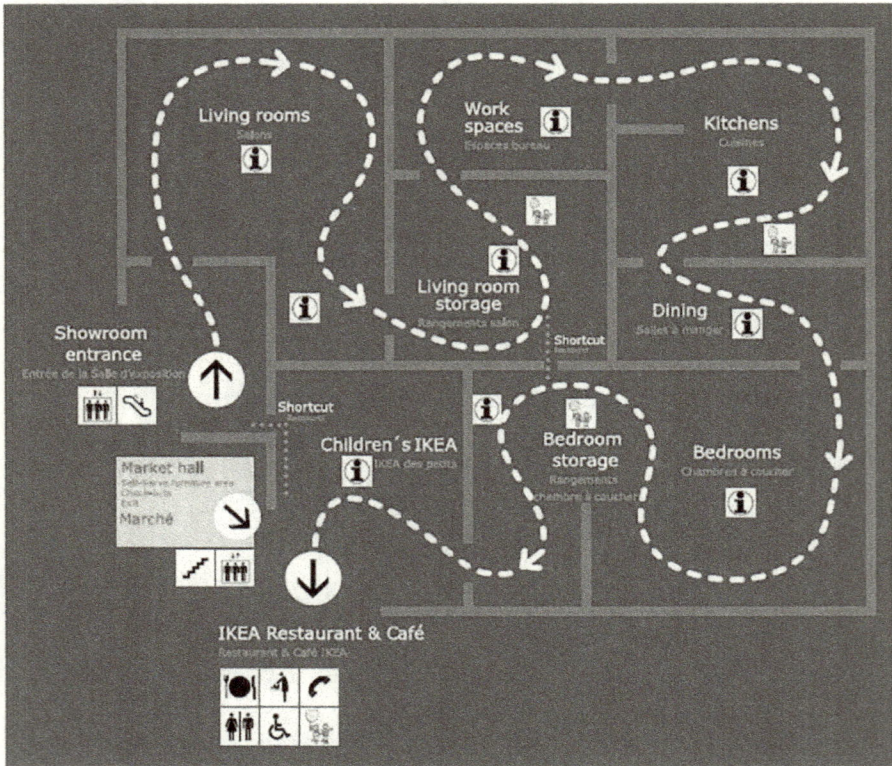

Our different shopping approaches highlight a crucial truth about user experiences: Not everyone wants to follow the same path to success. Imagine if IKEA forced everyone to wind through its showroom. It would frustrate focused shoppers like me while possibly overwhelming those who prefer to browse at their own pace. Instead, IKEA provides multiple ways to shop:

- Guided paths for browsers
- Direct routes for targeted purchases
- Online ordering for those who prefer to skip the store entirely

The same principle applies to B2B onboarding. As we've explored throughout this book, successful onboarding requires addressing three levels of friction: functional, social, and emotional. However, different users experience these friction points differently. For example, a technical founder migrating their startup's data faces different challenges than a marketing manager implementing a new workflow for their enterprise team.

This complexity makes personalization crucial for B2B onboarding success. Just as IKEA provides multiple paths to purchase, we need to create flexible onboarding experiences that adapt to different user needs and preferences. In this chapter, we'll explore how to personalize through customer jobs, skill levels, learning styles, job roles, job seniority, and business models. Let's start with the most fundamental approach: personalizing with Jobs-to-be-Done.

PERSONALIZING WITH JOBS-TO-BE-DONE

"Welcome to the product, {First_Name}!"

Many teams start and end their personalization efforts by inserting the user's name in the welcome modal or onboarding email. However, effective personalization goes beyond cosmetic changes. It fundamentally shapes the user's journey toward their Ultimate Win. When done right, it reduces all three levels of friction we explored in Chapter 2: functional friction through targeted guidance, emotional friction by addressing specific anxieties, and social friction by providing relevant organizational context and support.

The Jobs-to-be-Done (JTBD) framework, which we explored in Chapter 7, offers the most powerful approach to meaningful personalization. By understanding the different transformation goals users are "hiring" your product to achieve, you can create distinct paths that guide each user toward their specific version of success. This is particularly crucial for B2B products, where users often come with vastly different objectives and organizational contexts.

Like the bridge metaphor we explored in Chapter 1, each customer job requires its own bridge to success. Instead of building one generic bridge, you're creating multiple pathways—each designed to help users cross from their specific struggling moment to their desired transformation. Some users are crossing from manual bookkeeping to organized finances, while others are moving from unprofessional invoices to confident client communications. Each bridge needs its own support structures and clear markers of progress.

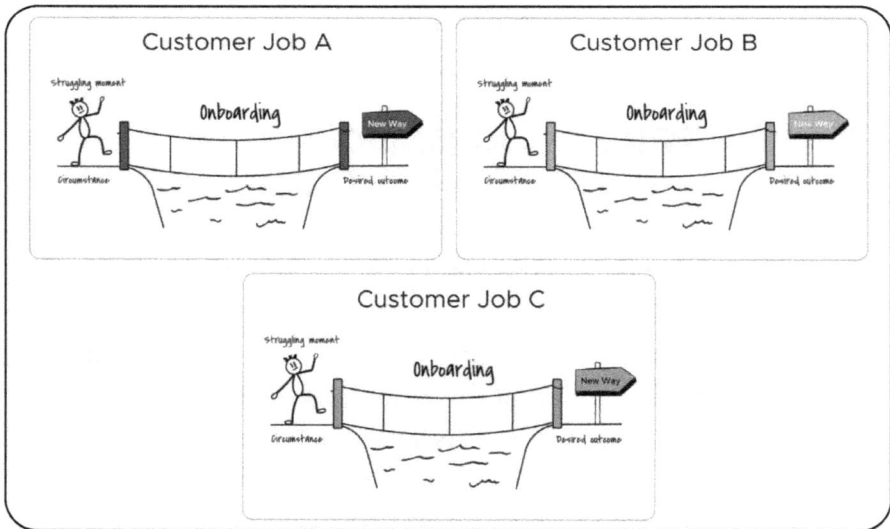

Wave Apps takes this onboarding approach. As we saw in Chapter 11, they discovered that trying to create a one-size-fits-all onboarding experience was failing their users. Through customer research, they identified three distinct customer jobs that required different onboarding paths:

1. Send Professional Invoices: Users trying to look more professional and get paid faster

2. Manage Business Accounting: Users struggling to stay compliant and organize finances

3. Run Employee Payroll: Users aiming to pay employees accurately and on time

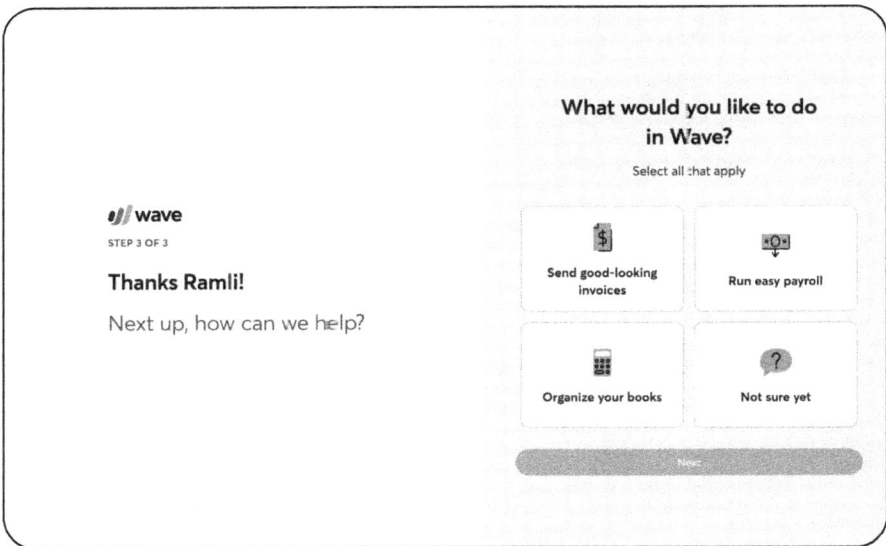

Each job demanded its own personalized experience:

- Different success metrics to track (e.g., time to first invoice vs. completed payroll run)

- Unique friction points to address (e.g., invoice design confidence vs. payroll compliance concerns)

- Tailored educational content (e.g., invoicing best practices vs. payroll tax guides)

- Specific product features to highlight

Instead of forcing all users through the same generic journey, Wave created branching paths based on the customer job selected during onboarding. This approach allowed them to show only relevant features and content for each job, address job-specific anxieties and concerns, and provide contextual help based on the user's goal. The result was a more focused, personalized experience that helped users achieve their specific objectives faster.

To implement this kind of jobs-based personalization, you can run through key exercises from the EUREKA framework for each customer job:

1. Complete the User Success Canvas and Four Forces of Progress (Chapter 9) to understand the specific transformation goals.

2. Run the Reverse Journey Mapping workshop (Chapter 12) to identify the Ultimate Win and map the Onboarding Success Journey.

3. Map the specific friction points using the Friction Mapping Exercise (Chapter 13).

4. Run the Friction-to-Action Workshop (Chapter 16) to identify targeted solutions.

5. Implement and measure improvements specific to that journey.

This process ensures that every aspect of your onboarding—from welcome messages to success metrics—aligns with the specific job your user is trying to accomplish. While this requires more upfront work than a generic onboarding flow, the improved activation rates and user satisfaction make it worthwhile.

After running through these exercises, you can create a high-level map showing the different paths of your onboarding journey. To help you visualize these branching journeys and how they interconnect, I've created the Personalized User Journey Template that you can download at *eurekabonus.com*.

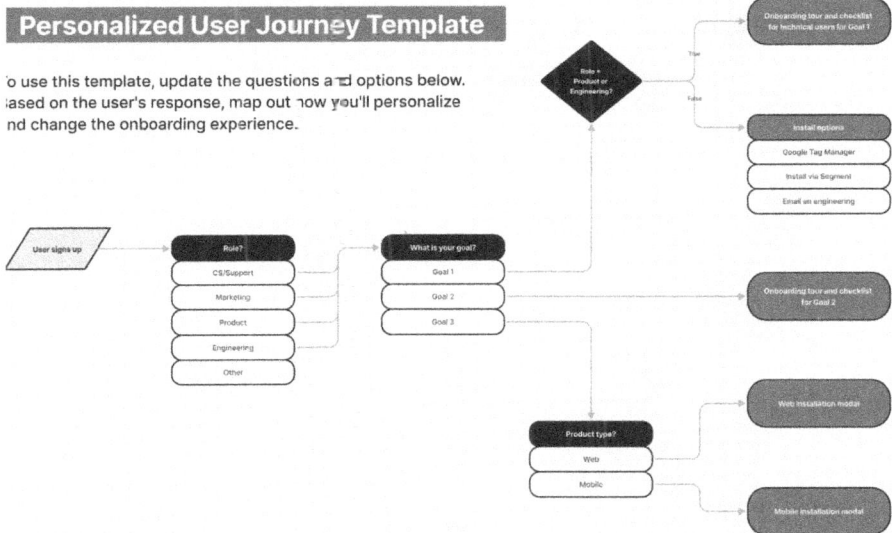

Personalized User Journey Template

To use this template, update the questions and options below. Based on the user's response, map out how you'll personalize and change the onboarding experience.

OTHER APPROACHES TO PERSONALIZATION

While segmenting by customer job is powerful, it's not the only way to personalize your onboarding experience. Here are three proven approaches:

1. Skill Level and Prior Knowledge

Products with technical components or industry-specific knowledge can benefit from adapting to users' existing expertise. Duolingo, for example, asks new users about their familiarity with their chosen language:

- "I'm new to this language"
- "I know some words and phrases"
- "I can have simple conversations"
- "I'm at an intermediate level or higher"

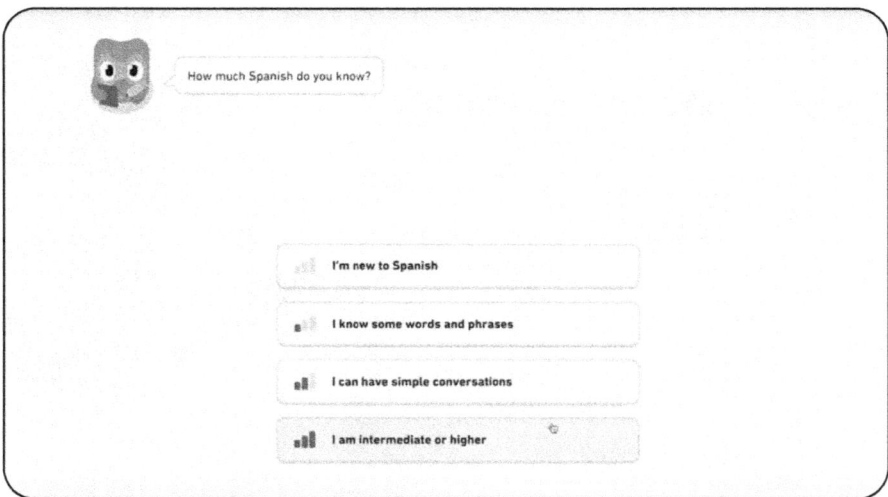

Based on the response, Duolingo either starts with basic lessons or offers a placement test.

HubSpot takes a similar approach. It asks users, "Have you used software to manage or engage with customers before?" Based on your response, you might start with basic CRM concepts like understanding a contact or with more advanced features like importing your customer list.

If you have a technical tool with a steep learning curve in a mature category (e.g., you have users switching from a direct competitor), personalizing based on skill level can help you cater to the onboarding experience so you don't overwhelm beginners or frustrate experienced users.

2. Learning Preferences

Another way to personalize your onboarding is to adapt to different learning preferences. Through years of user research and onboarding analysis, I've noticed that users typically fall into two learning styles: Clickers and Chillers.

Clickers are like dogs—enthusiastic explorers eagerly engaging with product tours, tutorials, and onboarding emails. Chillers are like cats—they prefer to learn and discover features at their own pace and often dismiss guided experiences.

Clickers   Chillers

Your product's target users often influence which style predominates. Marketing-focused tools like Canva tend to attract more Clickers, making interactive product tours and guided workflows effective. Developer tools like MongoDB or GitHub serve more Chillers, who prefer comprehensive documentation and community resources they can reference when needed.

Rather than guessing which type of user you are, smart products let users choose their learning style.

Crowd Content exemplifies this by offering users multiple onboarding paths:

- Watch a quick video overview

- Schedule a fifteen-minute guided demo

- Jump straight into the product

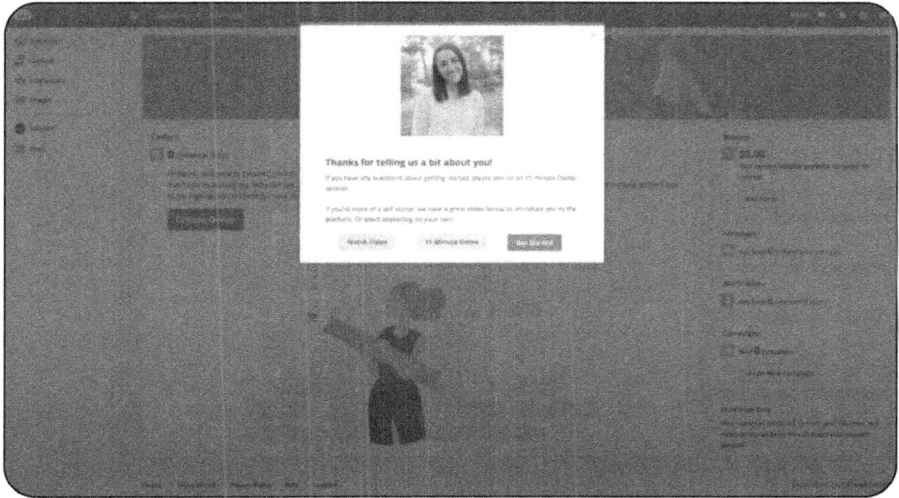

This flexible approach delivers three key benefits:

- Reduces friction by matching users' natural learning preferences

- Improves information retention through preferred learning methods

- Increases engagement by giving users control over their experience

If your product appeals to users with different skills or learning styles, offering multiple learning paths like Crowd Content lets users discover and absorb information in their preferred way.

3. Role and Responsibilities

Different roles within an organization often need different features and workflows. Fullstory exemplifies this approach by asking users about their roles after signing up.

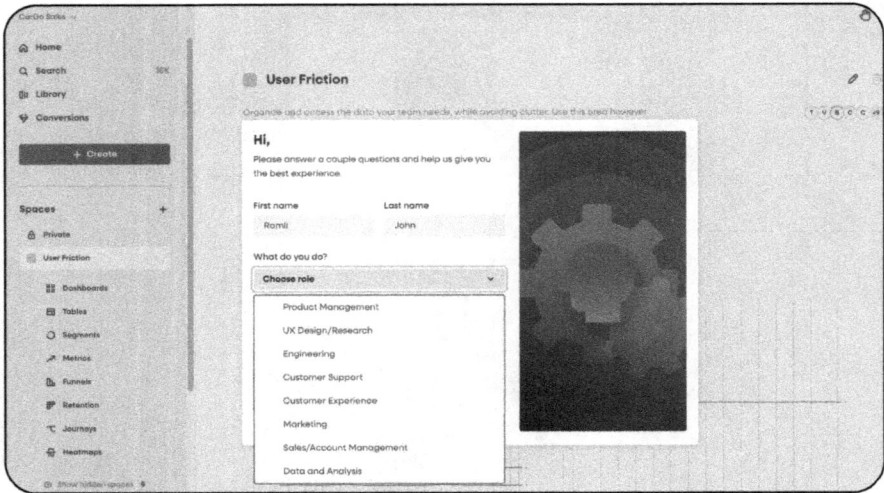

Then Fullstory personalizes their entire onboarding experience accordingly:

- Product Managers see UX-focused features like heat-maps and analytics dashboards for informing roadmap decisions.

- Engineers get direct access to dev tools and console logs for quick bug resolution.

- Support teams are guided toward session replay features for customer issue investigation.

- Data analysts receive workflows focused on metrics and segment creation.

This role-based personalization extends across all three pillars of their onboarding, with in-product guides highlighting relevant features for each role, email sequences containing role-specific use cases and tutorials, and educational content showcasing templates and workflows matched to each role's needs.

As Chrissy Quinones, Digital Customer Success Program Manager at Fullstory explains:

"Product managers care about visualizing UX patterns to prioritize their roadmap, while engineers need to quickly find and fix bugs. By understanding these different goals, we can guide each role to the features that matter most to them."

If your product becomes more valuable through team collaboration—whether for feedback, workflows, or project management—personalizing by job function helps users see how your product fits their specific role while understanding how it connects to their colleagues' work.

4. Job Seniority

Another powerful way to personalize onboarding is by adapting to users' seniority levels. Individual contributors often need hands-on guidance for daily tasks, while executives care more about team-wide implementation and ROI.

CloudApp (now Zight) demonstrates this approach effectively through their onboarding emails. Users who indicate they're an Account Executive receive an email focused on getting immediate value to close deals

"I noticed you are in a sales role at BuySellAds. Your customers will love you when you show them rather than telling them. The next time a customer asks, 'Can you show me that again?', wow them with a GIF or an annotated screenshot."

In contrast, emails to VP-level buyers focus on strategic value and team-wide implementation:

"Your time is valuable, and I wanted to help jumpstart the onboarding by offering a quick product overview to ensure your team gets the most out of it . . . Here are some quick resources on how CloudApp can help your sales, engineering, and customer team."

This seniority-based personalization recognizes that different levels need different types of support:

- Individual Contributors need practical how-tos and quick wins.

- Managers need team coordination tools and adoption strategies.

- Executives need ROI metrics and organizational impact.

By matching content and guidance to the seniority level, you help users achieve success within their specific scope of responsibility.

If buyers and end users sign up for your product, personalizing by seniority level lets you address each group's primary concerns, from individual productivity gains to organization-wide implementation and ROI.

5. Business Model

Different business models require different metrics and workflows. Amplitude demonstrates this approach by asking users about their business type during onboarding.

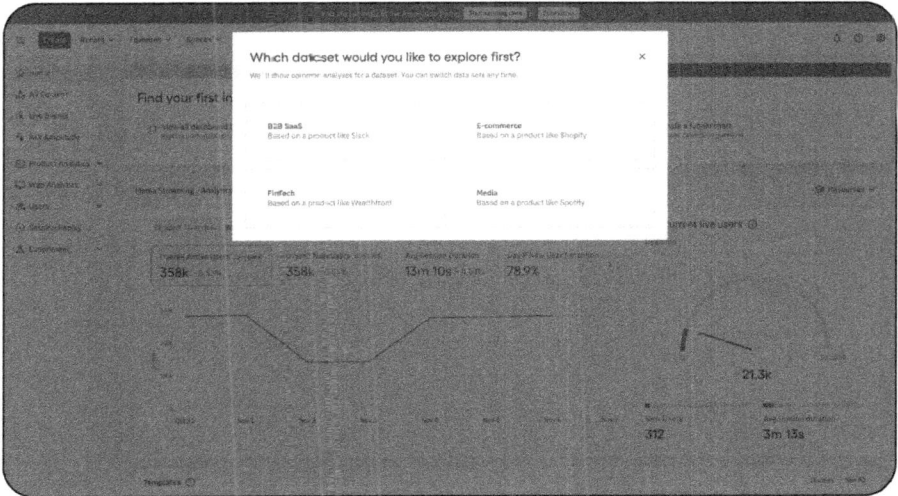

Then Amplitude personalizes their entire analytics experience accordingly:

- B2B SaaS companies see metrics like active users, activation rates, and Day Seven user retention.

- E-commerce businesses get dashboards showing average order value, cart conversion rates, and product page engagement.

- FinTech companies receive charts focused on transaction volume, user acquisition costs, and engagement patterns.

- Media companies see content engagement, subscriber retention, and advertising performance metrics.

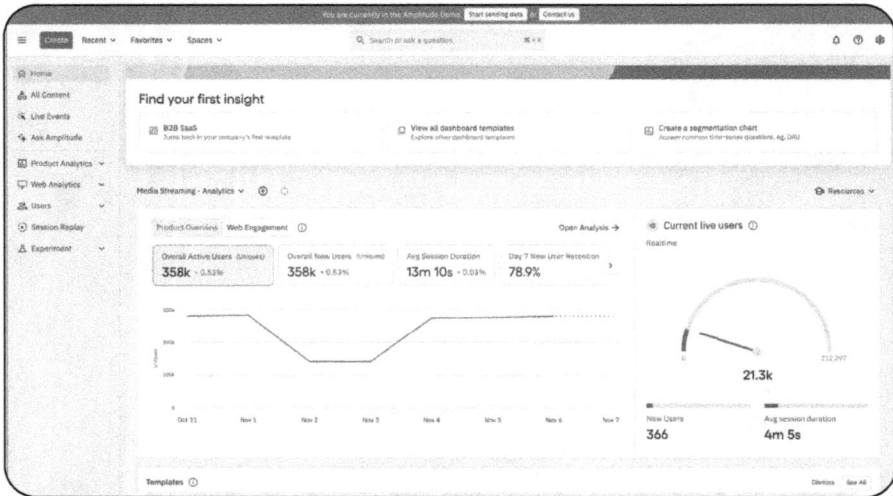

If your product serves multiple industries or business models, personalizing based on business type helps users immediately see relevant metrics and workflows without wading through irrelevant data.

CREATING YOUR PERSONALIZATION STRATEGY

Effective personalization isn't about implementing every possible approach—it's about choosing the right combination of questions that bring users closer to their Ultimate Win. Start with customer jobs as your foundation, then carefully select additional dimensions based on what will most impact their success:

- Product complexity (skill-level personalization)
- User diversity (role-based customization)
- Organizational structure (seniority-based approaches)

- Learning preferences (engagement style)
- Business model (industry-specific needs)

Remember: Every question you ask during signup should have a clear purpose in personalizing the user's journey. As a rule of thumb, limit yourself to three personalization questions—any more risks users dropping off during onboarding. Of course, test this with your specific product and users, but err on asking less and delivering more targeted value.

EUREKA ACTION ITEMS

Ready to personalize your onboarding? Here's where to start:

1. Review the customer jobs you identified in Step 1 of the EUREKA framework.

2. Go through the framework exercises for each job (User Success Canvas, Journey Map, etc.).

3. Use my Personalized User Journey Template at *eurekabonus.com* to map out other personalization strategies discussed in this chapter.

In the next chapter, we'll explore how to continuously improve your onboarding experience. You'll learn how successful teams run rapid experiments, measure what works, and iterate based on real user data.

18

Measuring and Analyzing Onboarding Success

~~~

What is not defined, cannot be measured.
What is not measured, cannot be improved.
What is not improved, is always degraded.

Lord Kelvin

~~~

"**What's our activation** rate?" asked the CEO during the quarterly review.

The room fell silent. Different teams shared conflicting numbers:

- Marketing reported email engagement rates.

- Product showed feature adoption metrics.

- Customer Success cited implementation milestones.

- Sales tracked trial conversion numbers.

This scene plays out in companies everywhere. Without clear metrics and systematic measurement, teams work at cross-purposes, each optimizing for different definitions of success.

In this chapter, we'll explore how to measure and analyze onboarding success across your entire customer journey. You'll learn how to use cohort analysis to validate improvements, build comprehensive dashboards that connect onboarding to business impact, and maintain momentum through continuous experimentation. Most importantly, you'll discover how successful teams use data to orchestrate improvements across all three pillars of onboarding—product guides, educational content, and human touchpoints.

COHORT ANALYSIS

Cohort analysis is the most effective way to measure your onboarding team's efforts. By comparing how different groups of users progress through your onboarding journey, you can clearly see the impact of your improvements. The simplest approach is creating two distinct cohorts:

- Users who experienced your original onboarding
- Users who experienced your improved onboarding

Cohort analysis helps you measure improvements across all three pillars of B2B onboarding success.

For in-product guides, compare users who complete your onboarding checklist against those who skip it. You might discover that users who finish the checklist reach their Ultimate Win 40 percent faster than those who don't. This data helps justify investments in better product tours and interactive walk-throughs.

With educational content, analyze the difference between users who engage with your academy or documentation versus those who don't. HubSpot consistently finds that users who complete their certification courses have significantly higher feature adoption rates and longer retention than non-certified users.

For human interaction, measure outcomes between customers who attend onboarding webinars or kickoff calls versus those who skip them. Salesforce data shows enterprise customers who participate in their guided onboarding program achieve ROI nearly twice as fast as those who don't.

For each comparison, focus on four key metrics:

- **Time-to-first-value**: The duration between signup and achieving their Same-Day Win. This indicates how quickly users get their first taste of success.

- **Time-to-full-value**: The duration between signup and reaching their Ultimate Win. This reveals how quickly users experience your product's core value.

- **Activation rate**: The percentage of new users who achieve their Ultimate Win within their first weeks. This indicates how effectively your onboarding guides users to success.

- **Day-N retention rate**: The percentage of users who return on day N (typically day seven, thirty, and ninety) after signup. This shows whether activated users develop lasting product habits.

- **Team-wide usage**: The percentage of invited team members who become active users. This reveals how well your product spreads within organizations.

InnerTrends's analysis below demonstrates the power of cohort comparison. The graph tracks daily retention rates between two groups: users who experienced the original onboarding versus those who went through an optimized version. The stark difference in retention curves validates the impact of their onboarding improvements.

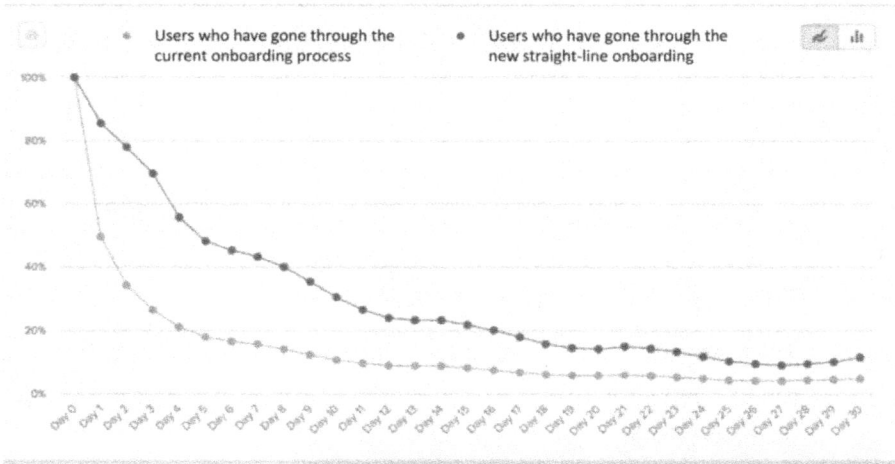

Let's examine two detailed examples of cohort analysis in action:

Case Study: Hotjar's A/B Test— Tooltips vs. Hotspots

Hotjar, a product experience insights platform, used cohort analysis to optimize their in-product guides. When introducing their "relevance selector" feature, they tested two approaches:

- Group A saw tooltip explanations
- Group B saw hotspots

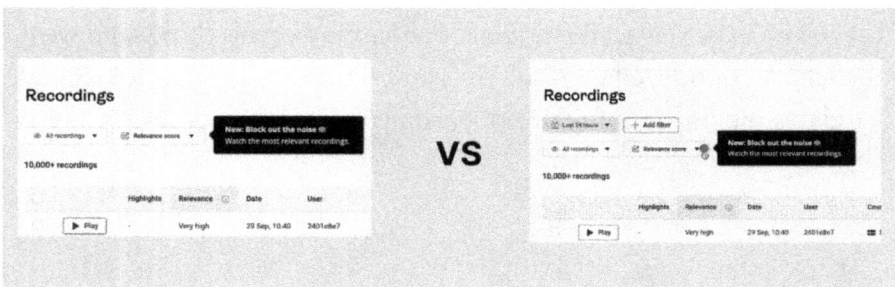

Both groups received identical content and copy, with users randomly split between variants. The results revealed clear differences:

- Tooltip cohort: 6.5 percent feature adoption rate

- Hotspot cohort: 8.16 percent feature adoption rate

- Overall improvement: 26 percent higher adoption with hotspots

Most importantly, the hotspot cohort showed 99 percent higher engagement with advanced features like saving recordings—demonstrating deeper product adoption beyond the initial interaction.

Case Study: HubSpot Academy's Data-Driven Approach

HubSpot Academy demonstrates how to measure educational content impact through cohort analysis. Instead of focusing solely on course completion rates, they compare:

- Activation rates between Academy graduates and non-participants

- Feature adoption rates among certified users versus non-certified

- Organization-wide adoption rates when admins complete certification

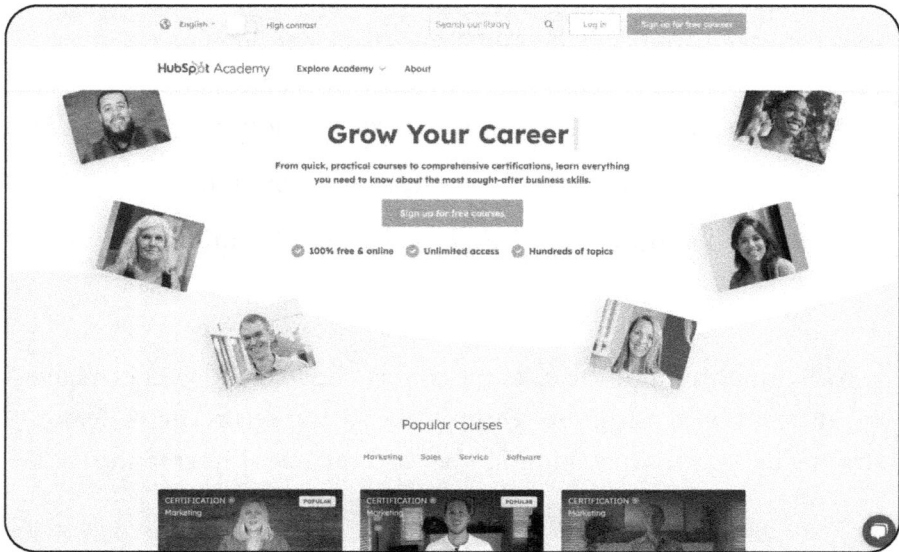

By connecting their Learning Management System with their CRM and product analytics, they can directly correlate educational engagement with customer success metrics like:

- Time to activation

- Feature adoption depth

- Net Promoter Score (NPS)

- Long-term retention rates

This data-driven approach helps them refine their educational content based on what demonstrably improves customer outcomes.

ONBOARDING METRICS BENCHMARKS

While it's tempting to compare your onboarding metrics against industry standards, you should treat these benchmarks with a healthy dose of skepticism. Every product is unique, and different user needs, value propositions, and complexity levels affect these metrics.

To give you a general sense of what other B2B companies achieve (see Appendix II for the sources of these benchmarks):

- Activation rates typically range from 25–35 percent for product-led B2B companies, while enterprise B2B products often see rates between 33–65 percent.

- Time-to-value averages around 1.5 days, though this varies significantly by industry (from one day for CRM tools to nearly four days for HR software).

- Free-to-paid conversion rates range from 3–8 percent for self-serve products to 10–15 percent for sales-assisted models.

- Week one retention often falls between 40–55 percent for Series A companies.

In **Appendix II: Onboarding Metrics Benchmarks**, I provide a more detailed breakdown and source of these benchmarks and other metrics across different business models and company stages.

Remember: Companies can artificially inflate these metrics by defining easier activation moments or measuring time-to-value from different starting points. Focus on improving your specific user journey rather than chasing industry averages that might not reflect your unique situation.

CREATING AN ONBOARDING DASHBOARD

Your onboarding metrics directly impact crucial business outcomes, from revenue growth and renewals to long-term retention. For product-led companies, effective onboarding often determines whether free users convert to paid plans. For sales-led organizations, it influences renewal rates and expansion opportunities. Thus, tracking onboarding success is essential for sustainable growth.

Ben Williams (PLG Adviser and founder of The Product-Led Geek) developed a comprehensive dashboard that connects onboarding success to business impact. While designed primarily for product-led companies, the dashboard's logical progression—from high-level business metrics to granular user actions—offers valuable insights for any B2B company. His framework, which you can find at PLG.news, organizes metrics into five key sections that sales-led businesses can adapt to their context:

1. **Business KPIs**: Core metrics tracking revenue, account plans, and daily activity to assess overall business health.

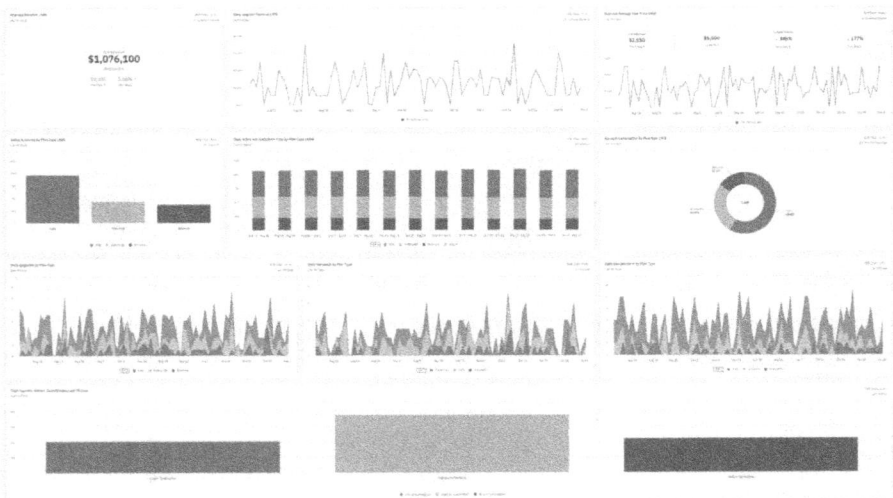

2. **Account Activation to Subscriber Conversion**: This feature tracks user progression from activation to paid conversion across multiple timeframes (1, 28, and 56 days).

3. **Time to Paid Conversion by Source**: Shows which acquisition channels convert fastest and most effectively.

4. **Churn by Acquisition Source**: Reveals which channels bring long-term customers versus high-churn risks.

5. **Monetization Trigger Conversion**: Maps specific actions that lead to upgrades or increased spending.

What makes this dashboard powerful is its ability to tell the complete story of your onboarding impact—from initial engagement through long-term retention. It connects individual user success to broader business metrics while providing actionable insights for improvement.

You can access Ben's complete Amplitude dashboard template at *amplitude.com/templates/product-led-geek-monetization-dashboard*.

To implement a similar dashboard for your product, focus on these core metrics:

- Time to Ultimate Win (by segment)
- Completion rates for key milestones
- Drop-off points in the journey
- Activation rates across cohorts
- The revenue impact of activated users
- Retention differences between activated/non-activated users

The key is ensuring these metrics flow between your essential tools—from analytics platforms like Amplitude or Mixpanel to your CRM, email tools, and customer success platforms. This integration enables you to orchestrate targeted interventions when users need support while demonstrating clear ROI from your onboarding investments.

FROM MEASUREMENT TO IMPROVEMENT

Having a comprehensive dashboard is only valuable if it drives action. The most successful teams use their metrics to fuel a continuous cycle of experimentation and improvement. I'm a big fan of the Build-Measure-Learn feedback loop popularized by Eric Ries in *The Lean Startup*.

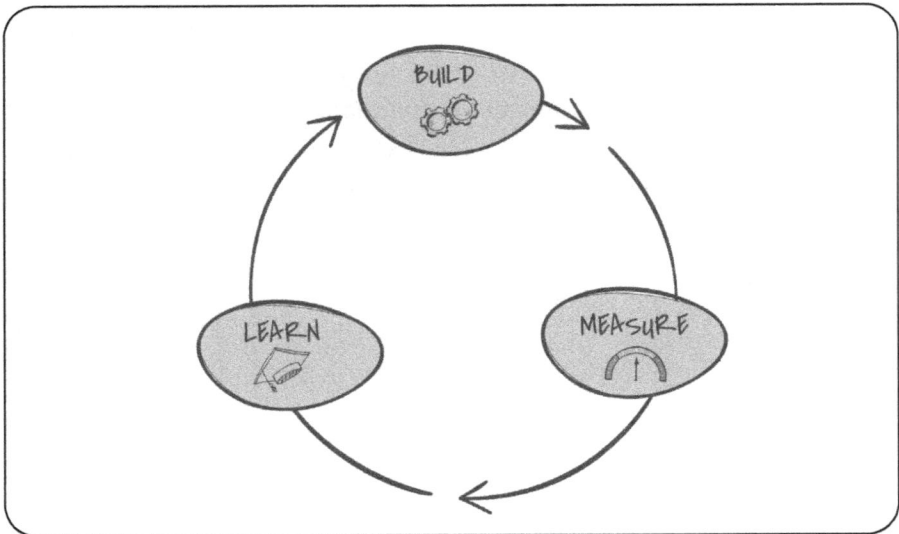

Instead of massive quarterly projects, run small, focused experiments based on your dashboard insights. When metrics show users struggling with a particular milestone, start with a minimum viable experiment:

- If completion rates drop at a specific step, test simplified instructions.

- When certain channels show poor activation, adjust their onboarding paths.

- If human-touch metrics indicate common questions, create targeted resources.

For example, rather than rebuilding your entire academy, test one focused video. Instead of overhauling your welcome flow, experiment with a single hotspot. Before launching full implementation programs, try a thirty-minute consultation call.

Here's how this iterative approach works in practice:

1. **Build (Week 1)**: Launch a small change, like reordering your checklist items based on usage patterns.

2. **Measure (Weeks 2–3)**: Track the impact through your dashboard metrics.

3. **Learn (Week 4)**: Analyze the results and plan your next experiment.

The key is maintaining momentum through regular reviews:

- Weekly: Monitor core metrics for sudden changes.

- Bi-weekly: Review experiment results and plan the next tests.

- Monthly: Analyze trends and adjust priorities.

- Quarterly: Assess the overall impact on business KPIs.

Remember: Four quick experiments often teach you more than one perfect solution. Use your dashboard to identify opportunities, but keep your experiments small and focused on moving specific metrics that matter.

SHARE WINS AND LEARNINGS

Most teams focus on running experiments but overlook a crucial multiplier—sharing learnings across the organization. Since onboarding touches every customer-facing team, visible progress builds company-wide momentum and support for continued experimentation.

Create regular channels to share your journey:

- Send a bi-weekly "Onboarding Wins" email highlighting experiment results.

- Use a dedicated Slack channel for quick updates and discussions.

- Host monthly lunch-and-learns to share team insights.

- Present quarterly reviews showing cumulative impact.

Focus on outcomes, not activities. Instead of "We launched a new checklist," share "Our latest experiment helped 25 percent more users reach their Ultimate Win." These outcome-focused updates accomplish three things:

- They build excitement across departments.

- They create empathy by showing real user impact.

- They justify continued investment in onboarding improvements.

Share both successes and failures openly. Failed experiments often teach the most valuable lessons, and discussing them builds a culture where teams feel safe to experiment. Frame updates around your User Success Canvas, showing how each improvement moves users closer to their transformation goals.

TURNING METRICS INTO MOMENTUM

Measuring onboarding success isn't about tracking metrics for metrics' sake—it's about creating a feedback loop that drives continuous improvement. Through cohort analysis, comprehensive dashboards, and regular experimentation, you can systematically identify what works and adapt your approach accordingly.

While industry benchmarks provide helpful context, remember that your metrics should reflect your unique user journey and business goals. Focus on collecting actionable data that helps you:

- Validate the impact of your onboarding improvements
- Identify where users need additional support
- Prove the ROI of your onboarding investments
- Guide your experimentation priorities

Most importantly, don't wait for perfect data or complete certainty. Start with the basics—tracking progress toward your Ultimate Win—and build from there. Small, measured improvements compound over time to create a significant impact.

In the next chapter, we'll explore what happens after users reach their Ultimate Win. You'll learn about "everboarding"—the ongoing journey of helping users discover deeper value and new use cases for your product.

NEXT STEPS
AND CONCLUSION

19

Everboarding: From Activation to Lifelong Product Adoption

~~~~~

Success is not a destination, it's a journey.

Zig Ziglar

~~~~~

When I first started rock climbing, I thought reaching the top of a route was the ultimate goal. But experienced climbers taught me something surprising: The real achievement isn't just reaching the summit once—it's being able to climb increasingly challenging routes and helping others discover the joy of climbing.

The same principle applies to B2B product onboarding. Many companies celebrate when users reach their Ultimate Win, thinking their job is done. They treat onboarding like a sprint to the finish line rather than the start of a longer journey. But just as reaching the top of one climbing route reveals new challenges ahead, achieving the Ultimate Win opens doors to deeper product value and wider use cases.

I saw this principle in action at Appcues. While publishing a first flow was our Ultimate Win, our most successful customers went far beyond—creating flows for feature announcements, NPS surveys, and more. Each new use case needed its own guidance and best practices, showing us that onboarding truly was a continuous journey, not a destination.

This continuous learning approach has a name: **Everboarding**. I first heard the term from Courtney Sembler, Senior Director of HubSpot Academy. It recognizes that mastery isn't a destination but an ongoing journey of discovery. In this chapter, we'll explore how to apply everboarding principles to product success through:

- The two paths of everboarding: going deeper vs. wider
- Choosing the right strategy
- Implementation and measurement

Let's start by understanding why treating onboarding as a one-time event limits product success.

WHY EVERBOARDING MATTERS

Remember World 1-1 in *Super Mario Bros.*? That first level masterfully teaches players the basic mechanics—running, jumping, and dealing with enemies. But completing World 1-1 isn't the end of Mario's journey; it's just the beginning. Each new level introduces more complex challenges and mechanics, following the gaming principle of "new levels, new devils."

Just as Mario faces increasingly complex challenges in later levels, your users need guidance to master advanced features and unlock new value. Let's explore three crucial reasons why that's important.

1. Everboarding Reduces Churn Through Deep Adoption

Everboarding creates a powerful defense against churn by embedding your product deeply into users' daily work. Think of it like roots

3. **New Problems Create Cross-Sell Opportunities:** Teams mastering one product naturally need others—like HubSpot users expanding from Marketing to Sales Hub. It opens up the door to sell your other products.

3. Everboarding Creates Product Champions

Advanced users naturally become your most valuable advocates, as seen with Notion's power users. Starting with basic docs and wikis, they progress to building sophisticated systems with databases and formulas, transforming how their teams work. Inside their organizations, they become productivity experts—creating templates, running workshops, and building custom workspaces that drive adoption across departments. Outside their companies, they share their expertise through social media, detailed guides, and active user communities.

These champions also shape your product's future in powerful ways. Their deep understanding of both your product and real-world business challenges leads to highly practical feature requests—like the user feedback that influenced Notion's AI features. Their authentic enthusiasm and creative solutions inspire others through viral social media content, workspace tours, and community contributions, creating a network effect more powerful than traditional marketing could ever achieve.

THE TWO PATHS OF EVERBOARDING

After users achieve their Ultimate Win, successful B2B products guide them down one of two paths: going deeper into existing capabilities or wider into new use cases. Let me illustrate both paths through HubSpot's journey with their customers.

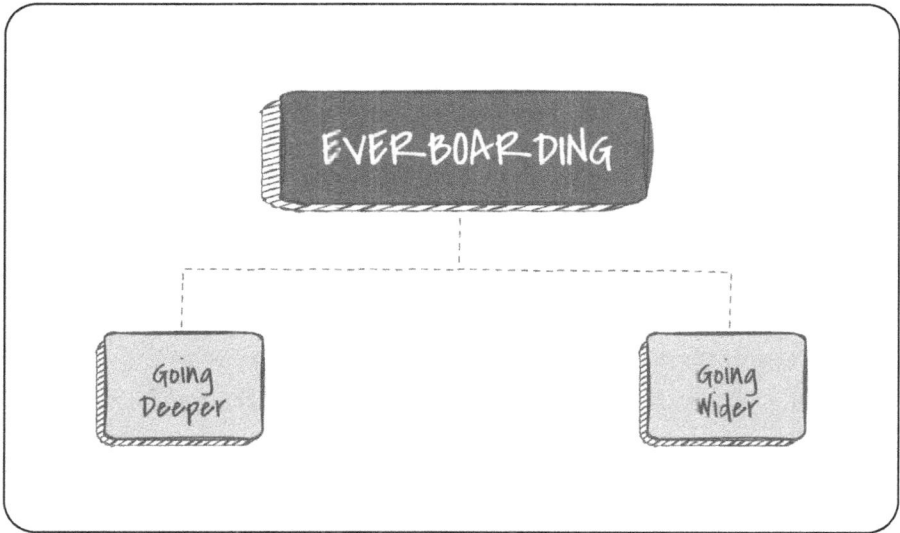

Going Deeper

Many marketing teams start with HubSpot's Marketing Hub to create basic landing pages and email campaigns. After generating their first qualified leads (their Ultimate Win), some teams want to master digital marketing more deeply. They progress from basic lead generation to sophisticated inbound marketing strategies.

The deeper path helps users become more advanced with Marketing Hub. A marketing team might advance to:

- Creating dynamic content personalized to visitor segments
- Building sophisticated marketing automation workflows
- Implementing SEO strategies with content pillar pages
- Setting up advanced lead scoring and qualification rules
- Developing multi-channel campaign attribution

Each advancement helps them become more effective marketers. They're not tackling new departments' challenges; they're gaining more sophistication with their marketing capabilities.

Going Wider

Other customers choose to expand across HubSpot's entire platform. After mastering Marketing Hub, they discover how HubSpot can connect their entire customer journey. The expansion typically flows naturally:

- Sales teams adopt Sales Hub to align with marketing's qualified leads.

- Service teams implement Service Hub to support converted customers.

- Operations teams deploy Operations Hub to maintain clean, connected data.

- CMS Hub powers their website with native marketing tools.

Each expansion creates a more complete view of the customer journey. While teams can apply some knowledge between hubs, each product requires its own learning journey. The result? A unified platform where Marketing attracts leads, Sales converts them efficiently, and Service teams deliver exceptional support—all while sharing the same customer data and insights.

The choice between going deeper or wider often depends on team priorities and organizational needs. Some companies need to perfect their marketing operations before expanding to sales and service. Others find more value in connecting their entire customer journey immediately. The key is identifying which path aligns with your users' growth trajectory and supporting them accordingly.

CHOOSING YOUR EVERBOARDING STRATEGY

Successful everboarding requires careful consideration of which path—deeper or wider—will deliver the most value for your users. While every company's journey differs, three key factors often influence this decision.

1. User Maturity

Start by analyzing your most successful customers' journeys after their Ultimate Win. Watch for natural progression patterns in feature adoption and team expansion. Your power users often reveal the most logical paths for growth, showing you where others might naturally advance.

2. Pricing Model

Your pricing structure naturally influences expansion strategy. Per-seat pricing models like Slack benefit from wider team adoption, while feature-based tiers like HubSpot's Hubs encourage deeper usage within each product before cross-selling. Let your pricing guide users toward their most valuable next steps.

3. Customer Readiness

Most importantly, match your everboarding to your customers' organizational readiness. A marketing team struggling with basic lead generation needs to master those fundamentals before exploring automation. However, if they're experiencing friction with their sales team, expanding into Sales Hub might deliver more immediate value.

Remember: The most effective everboarding path aligns your product's capabilities with your customers' natural growth trajectory. Focus on making each progression feel like an obvious next step rather than a forced expansion.

THE EUREKA FRAMEWORK FOR EVERBOARDING

The EUREKA Framework we explored throughout this book adapts naturally to everboarding. You can apply each step to support customers beyond their Ultimate Win:

1. Establish Your Everboarding Team

While your initial onboarding team focused on activation, your everboarding team should include members focused on expansion and growth. Product marketing helps identify new use cases, Customer Success drives deeper adoption, and Sales enables cross-product expansion. This cross-functional collaboration ensures users continue discovering new value.

2. Understand Extended User Success

Success looks different after the Ultimate Win. Map out what advanced success means for different user types. A power user might aim for workflow automation mastery, while a team admin might focus on driving organization-wide adoption. Understanding these advanced success states helps you guide users toward their next achievements.

3. Reverse Journey Map Advanced Milestones

Just as we mapped the path to the Ultimate Win, chart the journey to advanced milestones and wins. Start with sophisticated use cases and work backward to identify the stepping stones users need. For example, before a HubSpot user can implement complex marketing automation, they need to master audience segmentation, email workflows, and conversion tracking.

4. Keep Users Engaged with Advanced Features

Just as we mapped friction points in initial onboarding, advanced users face new barriers to adoption. Map these friction points across three levels:

Functional friction now involves complex technical challenges like API integrations, custom workflows, and advanced configurations. Address these through targeted in-product guides that break down sophisticated features into manageable steps.

Social friction shifts to organizational change management and cross-team coordination. Support this with educational content like implementation playbooks and certification programs that help champions drive adoption across their organization.

Emotional friction centers on confidence with advanced features and trust in expanding usage. Overcome these barriers through strategic human touchpoints, including advanced training sessions and expansion planning meetings.

Apply these insights through the three pillars of engagement:

- Use in-product guides to announce new features and showcase advanced capabilities.
- Create advanced educational content like power-user certification programs.
- Provide strategic human touchpoints to guide enterprise customers toward expansion.

5. Apply, Analyze, and Repeat

Measure the success of your everboarding efforts through:

- Advanced feature adoption rates
- Cross-product expansion metrics
- Team-wide usage growth
- Customer lifetime value increase
- Account expansion revenue.

Use these insights to refine your everboarding approach continuously. Remember: Just as initial onboarding never truly ends, everboarding constantly evolves to help users extract more value from your product.

THE JOURNEY CONTINUES

Like reaching the summit of rock climbing, your product's Ultimate Win isn't the end—it's a launchpad for deeper exploration. Throughout this chapter, we've seen how everboarding reduces churn, expands revenue, and creates passionate champions through two key paths: going deeper into existing features or wider into new use cases. Companies like HubSpot, Asana, and Notion show us how this works.

The EUREKA Framework guides this journey, helping you build the right team, map advanced success milestones, and create engagement strategies that evolve with user expertise. Through everboarding, you transform your product from a tool into an essential part of users' daily work, creating a network of value that becomes increasingly difficult to replace. This is how great B2B products turn activation into lifelong adoption.

Your Turn: Map Your Everboarding Journey

- Map out what success looks like beyond your Ultimate Win.
- Identify opportunities for users to go deeper into existing features.
- List potential expansion paths for different user segments.
- Design guidance for key advanced features.
- Plan how you'll measure ongoing user growth.

CONCLUSION

Bridging the Onboarding Journey

I will not tell you how long or short the way will be; only that it lies across a river. But do not fear that, for I am the great Bridge Builder.

Aslan, The Voyage of Dawn Treader, C.S. Lewis

One summer, my wife Joanna and I took our family hiking near my parent's house in Toronto. The trail promised a scenic route through one of the city's many wooded ravines—if we could reach it. Between us and our destination stood a weathered wooden bridge, its planks suspended over a steep drop into the ravine below.

As we approached, our excitement turned to apprehension. Some planks were missing entirely, and others looked rotten. The railings, weathered and worn, swayed with each gust of wind. Each step would need careful testing before we committed our full weight.

I went first, gingerly testing each plank.

"This one's solid," I called back.

Step by step, plank by plank, we made our way across. It took three times longer than a sturdy bridge would have required, but eventually, our whole family reached the other side safely. The view of the ravine's beauty was breathtaking, but the anxiety of that crossing dampened everyone's initial excitement.

Unfortunately, this is exactly how many B2B companies approach onboarding. They leave customers to test each step cautiously, wondering if it will support their journey to success. Some planks are missing—key features unexplained or crucial guidance absent. Others are rotted through—outdated documentation or broken workflows. The whole experience feels precarious, requiring constant vigilance instead of confident progress.

It doesn't have to be this way.

Throughout this book, we've explored how to build sturdy bridges that inspire confidence rather than anxiety. The EUREKA Framework provides the blueprint for creating onboarding experiences that safely and efficiently guide users to their desired transformation:

ESTABLISH YOUR ONBOARDING TEAM

Just as building a bridge requires architects, engineers, and construction crews working in harmony, successful B2B onboarding demands cross-functional coordination. We learned how to align Product, Marketing, Sales, and Customer Success teams around shared goals and clear ownership.

UNDERSTAND USER SUCCESS

Before constructing any bridge, you must know exactly where people need to go. Through the User Success Canvas and Four Forces analysis, we discovered how to deeply understand the transformation users seek—not just functionally but emotionally and socially as well.

REVERSE JOURNEY MAP TO SUCCESS

Like engineers who design bridges by starting with the destination, we learned to work backward from the Ultimate Win to create clear pathways to success. By mapping Same-Day Wins and key milestones, we ensure users maintain momentum throughout their journey.

KEEP NEW CUSTOMERS ENGAGED

The strongest bridges have multiple support structures. Through the three pillars of successful onboarding—in-product guides, educational content, and human touchpoints—we explored how to provide comprehensive support that keeps users moving forward.

APPLY, ANALYZE, AND REPEAT

Just as bridges require regular inspection and maintenance, onboarding needs continuous measurement and improvement. We learned to run rapid experiments, measure their impact, and create feedback loops that drive ongoing optimization.

The journey doesn't end when users reach their Ultimate Win. Like a bridge that enables daily commutes, deliveries, and connections, successful onboarding creates lasting transformation. Through everboarding, we help users discover new destinations, unlock deeper value, and become champions who guide others across the bridge.

Remember: You're not just implementing software—you're enabling transformation. Every improvement to your onboarding experience helps more users successfully cross the chasm between struggle and success. While the journey may seem daunting, the EUREKA Framework provides a proven blueprint for building lasting bridges.

Now it's your turn to become the Bridge Builder. Here's where you start:

1. **Inspect your current bridge**
   - Gather your cross-functional team.
   - Run the Sailboat Exercise.
   - Map out where users currently struggle.

2. **Design your improved crossing**
   - Define your Ultimate Win.
   - Identify your Same-Day Win.
   - Create clear pathways between these milestones.

3. **Build your support structures**
   - Implement targeted in-product guides.
   - Create valuable educational content.
   - Deploy strategic human touchpoints.

4. **Maintain and strengthen your bridge**
   - Track progress toward key milestones.
   - Run rapid experiments.
   - Share learnings across teams.

Most importantly, remember that onboarding isn't a one-time construction project—it's an ongoing commitment to helping users reach their destination safely and confidently. Each improvement strengthens the bridge that enables their transformation.

Let me leave you with this parting thought:

Like Aslan in C.S. Lewis's tale, we must become great Bridge Builders ourselves—creating safe passages that guide users and their teams from their struggling circumstances to their desired transformation.

The journey may be long or short, but with the right framework and tools, we can build bridges that stand the test of time.

Onward together,
Ramli John
Ramli "RJ" John

PS—Thank you for joining me on this journey. I'd love to hear how you implement the EUREKA Framework at your company. Email me at ramli@delightpath.com or connect with me on LinkedIn (linkedin.com/in/ramlijohn).

PPS—If you found this book helpful, please consider leaving a review at eurekabook.co/review. Your feedback helps other B2B teams discover these frameworks and join our mission of building better bridges to customer success.

growing deeper and wider—the product becomes the organization's central nervous system. Users don't just stick around because switching is difficult; they stay because the product delivers compounding value they can't afford to lose. Even if competitors offer lower prices, the cost of migrating years of workflows and retraining teams outweighs any potential savings.

Consider Asana: A marketing team starts with basic content calendar management and progresses to advanced features like custom fields, forms, and automation. As they integrate with tools like Slack and Google Drive, the product becomes essential to their workflow. Soon, other teams see this success and adopt Asana for their own needs, each adding unique workflows and templates.

2. Everboarding Drives Revenue Expansion

Advanced users naturally spend more money on your product. HubSpot exemplifies this growth pattern perfectly: Marketing teams often start with basic email campaigns in Marketing Hub. As they master core features, they unlock advanced capabilities like workflow automation and personalization. Their visible success attracts attention from the sales team, who want similar results for their processes, which drives the adoption of HubSpot's Sales Hub.

Companies expand their revenue contribution in three key ways:

1. **Deeper Usage Drives Lifetime Value:** As users master advanced capabilities, they solve bigger problems and invest more in your platform.

2. **Team Growth Increases Seats:** Success spreads—like Slack growing from five marketing users to fifty across multiple departments. Each new team adds recurring revenue through per-user pricing models.

ACKNOWLEDGMENTS

Writing this book has been both rewarding and challenging. As someone with a math degree who learned English as a second language, crafting these pages stretched me far beyond my comfort zone. Without the encouragement and feedback from the people below, this book wouldn't exist.

Joanna: My wife, my soul mate, my best friend. You are my sunlight and God's favor in my life. You nursed me back to health when I got sick halfway through writing this book. You encouraged me and served as a sounding board during the ups and downs of the writing process. Your endless affection forever leaves me in your debt; I'm wondering how I got so lucky. Here's to many more years together!

Zane: My life changed when I first held you in my arms. Watching you grow into a compassionate, curious, and confident toddler makes me proud to be your dad. One day, you'll reach the stars and achieve better and greater things than Mama and Daddy.

Ramon and Lina: My parents, my foundation. You always encouraged me to pursue big dreams and never settle for less, rooting my faith in God. Your love and pride have been constant forces in my life, shaping who I am today. (Fun fact: My name, "Ramli," comes from combining their names—Ram-Li—a perfect blend of the two people who taught me to dream big.)

Ben Putano: As my first book as a solo consultant, I couldn't have found a better guide through the publishing process. Your expertise at Damn Gravity helped transform my ideas into reality, making this journey far less daunting than it could have been.

Wes Bush: My mentor and friend from ProductLed, who helped refine the EUREKA Framework in our first book together, *Product-Led Onboarding*. You saw the potential in my ideas when they were just rough notes. Thank you for believing in this vision from the start.

Vivek Balasubramanian: Your expertise, having led product growth at companies like Roofr, Shopify, and Wave, made your detailed review invaluable. Thank you for having the courage to give such honest and thorough feedback, which helped sharpen the book's focus on what truly matters for B2B onboarding success.

Dave Rigotti: Thank you for the multiple discussions about the complexities of B2B onboarding, or as you aptly call it, account-based onboarding. Your insights were invaluable.

Chrissy Quoines: Your work at Fullstory and willingness to share your story and process added a crucial real-world perspective to this book.

Courtney Sembler: Our conversation helped save my chapter about everboarding. I first heard that term from you. Thank you for being an inspiration for all your work at HubSpot Academy!

Auroriele Hans: From joining my cohort course to providing invaluable feedback in the early reader's club, your enthusiastic support has helped shape this book. Thank you!

Stephanie Lauderback: You not only brought the Delight Path brand to life, but this book is immeasurably better thanks to your beautiful cover design and illustrations. Thank you for helping readers better understand these concepts through your art.

Karim Hussain, Akash Mahajan, Sean McCarthy, Morgan Smith, Tom Crossman, Patricio Hervas, Jos van der Jooji, Courtney Semblar, Marc Thomas, Giuliana Hejtmanek, and Dave Rigotti: Thank you for being early readers and providing feedback when this book was barely written.

Axel Sukianto, Credic Habex, Rob Heiderman, Pradip Khakhar, Fracois Simichiev, Ahmed Darwish, Benjamin Fourio, Clay Ostrom, Jimmy Louchart, Darren Fantor, Auroriele Hans, Timothy Bryan Berry, Gerard Dolan, Amira Abo Doh, Erik Macarney, Nirhidayati Aziz, Ali Rushdan Tariw, Warwick Eade, Walid Koleilat, Kasia Flood, Elizabeth Johnson, Shannon Howard, Georgiana Lauid, Michel Hauzer, Amy M, Matt Tidwell, Ahmed Magdi, Sangeetha Iyer, William Frimel, Luisa Siliprandi, Rajat Chadda, Ralitsa Minkova, Diego Garea Rey, Kumar Utsav, Kevin Struthers, Chandra Manubothu, Kay Del Rosario, Krystal Motwani, Koranz Stoler, Philippe Tutula, Fabine Fentker, and David Jenyns: Thank you for pre-ordering the book before I've even released the first draft.

Most importantly, all my students and clients: Thank you for allowing me to serve you.

ABOUT THE AUTHOR

Ramli "RJ" John is the founder of Delight Path and author of the bestselling book *Product-Led Onboarding*. With 40,000 copies sold worldwide, his book has become the go-to guide for thousands of B2B leaders looking to build an onboarding strategy that turns their users into lifelong customers. RJ's worked with companies such as Zapier, Leadpages, Appcues, Vidyard, Bynder, and other fast-growing B2B companies to level up their onboarding experience. On a personal note, he lives in Toronto, Canada, with Joanna (his wife), Zane (three-year-old toddler), and Ro (six-year-old dog).

APPENDIX

APPENDIX I

Using Artificial Intelligence to Improve Your Onboarding

Artificial intelligence is transforming how we build and optimize onboarding experiences. While the technology evolves rapidly—making specific tool recommendations quickly outdated—three key applications of AI show particular promise for improving B2B onboarding:

1. SYNTHESIZING USER RESEARCH AND TEAM INSIGHTS

One of the biggest challenges in improving onboarding is making sense of vast amounts of qualitative data from customer interactions. AI can help by analyzing:

- Customer onboarding call recordings
- Sales conversation transcripts
- Support ticket patterns
- User feedback surveys
- Team meeting notes

Tools like NotebookLM, Claude, or other AI agents can process these inputs to identify common friction points, surface recurring user questions, and highlight successful onboarding patterns.

This analysis helps teams spot patterns they might miss manually, leading to more targeted improvements in their onboarding experience. You can find a detailed example of how I use NotebookLM to analyze customer interviews and synthesize insights at *eurekabonus.com*, including my prompt templates and analysis workflow.

2. CREATING EDUCATIONAL CONTENT AND COMMUNICATIONS

AI writing assistants can help teams create more effective onboarding materials by:

- Generating first drafts of onboarding emails

- Suggesting improvements to help documentation

- Creating variations of product tours for different user segments

- Adapting technical content for different audience levels

- Maintaining a consistent voice across materials

The key is using AI as a collaborative partner—having it generate initial drafts that human editors then refine for accuracy and brand voice. At *eurekabonus.com*, you'll find a walkthrough and resources showing how the Lex AI team used their own AI tools to craft and optimize their onboarding email sequences, demonstrating this collaborative approach in action.

3. PERSONALIZING THE ONBOARDING EXPERIENCE

Perhaps the most exciting application of AI is creating dynamically personalized onboarding experiences. AI can analyze user behavior patterns and characteristics to:

- Adjust onboarding paths based on user roles and goals

- Recommend relevant features and resources

- Trigger contextual help at optimal moments

- Predict when users might need human support

- Customize success celebrations

While this level of personalization is still emerging, early experiments show promise in creating more relevant, engaging onboarding journeys.

THE FUTURE OF AI IN ONBOARDING

The landscape of AI tools and capabilities changes weekly. What seems cutting-edge today might be standard practice tomorrow. Rather than focusing on specific tools, consider how AI might help you:

- Process more user feedback faster

- Create more targeted content efficiently

- Deliver more personalized experiences

- Identify improvement opportunities earlier

- Scale human-like interactions

For the latest AI tools and implementation strategies, visit *eurekabonus.com*. As technology evolves, I'll regularly update this resource with new developments and practical applications.

Remember: AI should augment, not replace, human judgment in onboarding design. Use it to handle repetitive tasks and surface insights, freeing your team to focus on strategic decisions and human connections that truly drive successful onboarding.

APPENDIX II

Product Onboarding Benchmarks

While benchmarks can provide helpful context, they should be taken with a grain of salt. Every product is unique, with different user needs, value propositions, and complexity levels that affect these metrics. Additionally, how companies define and measure these metrics varies significantly.

1. ACTIVATION RATES

The activation rate is calculated as:

**Activation Rate = [Users who hit your activation milestone] /
[Users who completed your signup flow]**

Your activation milestone (often called the "aha moment" or what I refer to as the Ultimate Win in this book) is the earliest point in your onboarding flow that demonstrates your product's value and predicts long-term retention. Users are "activated" when they reach this milestone.

Here are the activation rate benchmarks across different product types and go-to-market strategies:

Here are some average activation rate benchmarks that Lenny Rachitsky shared in his newsletter (lennysnewsletter.com) after surveying five hundred companies:

- B2B SaaS average: 36%

- B2B SaaS median: 30%

B2B Enterprise SaaS (e.g., Salesforce, Workday):

- Median: 33%

- 60th percentile: 40%

- 80th percentile: 65%

B2B Product-Led SaaS (e.g., Figma, Canva):

- Median: 25%

- 60th percentile: 30%

- 80th percentile: 46%

Take these activation rate benchmarks with some healthy skepticism. Companies can artificially inflate them by defining easier activation moments. For instance, choosing "adding the first contact" instead of "creating their first accurate pipeline forecast" as the Ultimate Win will show higher activation rates but may not reflect true user success or predict long-term retention.

2. TIME-TO-VALUE (TTV)

Time-to-Value (TTV) is the duration it takes for a new user to experience the core value of a product after starting to use it. It measures how quickly a product can deliver its promise to users, indicating the efficiency and effectiveness of the onboarding process.

Here are some TTV benchmarks from the UserPilot team found after analyzing 327 companies:

Overall Industry Averages:

- Average: 1 day, 12 hours, 23 minutes

- Median: 1 day, 1 hour, 54 minutes

- Standard Deviation: 1 day, 9 hours, 28 minutes

Here are the benchmarks across different dimensions:

By Company Size (Annual Revenue):

- $1M - $5M: ~1 day, 5 hours

- $5M - $10M: ~1 day, 7 hours

- $10M - $50M: ~2 days

- $50M+: ~1 day, 16 hours

By Industry:

- HR: 3 days, 19 hours

- Martech: 1 day, 21 hours

- Fintech and Insurance: 1 day, 17 hours

- Healthcare: 1 day, 7 hours

- CRM & Sales: 1 day, 5 hours
- AI & ML: 1 day, 17 hours

By Growth Strategy:

- Product-Led Growth (PLG): 1 day, 12 hours
- Sales-Led Growth (SLG): 1 day, 11 hours

Similar to activation rates, I'm cautious about TTV benchmarks. Companies can artificially reduce their time-to-value by defining an earlier, easier-to-achieve moment as their measure of "value delivered."

3. FREE-TO-PAID CONVERSION

Free-to-paid conversion measures the percentage of new accounts that become paying customers within their first six months. It's calculated as:

**Free-to-paid conversion =
[new accounts who begin paying within their first 6 months] /
[total new accounts created during the measurement window]**

The most effective way to analyze this metric is through cohort analysis, tracking how different groups of users convert over time. This approach helps you understand how changes to your onboarding, product features, or pricing affect conversion rates for specific user segments.

Based on Lenny Rachitsky's (LennysNewsletter.com) survey of product-led companies, here are the conversion rate benchmarks across different business models:

Self-Serve Freemium:

- Good: 3–5%
- Great: 6–8%

Sales-Assisted Freemium:

- Good: 5–7%
- Great: 10–15%

Free Trial Products:

- Good: 8–12%
- Great: 15–25%

Free trial products typically see higher conversion rates as users often sign up when they're more ready to buy.

4. WEEK 1 RETENTION

Week 1 retention measures the percentage of new users who return to your product within 7 days of signup. It's calculated as:

Week 1 Retention =
[Users who returned within 7 days of signup] /
[Total users who signed up]

According to June.so's analysis of SaaS companies in their product benchmark report, here are the Week 1 retention benchmarks by company stage:

Pre-seed Stage:

- Top quartile (75th percentile): >51%
- Median (50th percentile): 43%
- Bottom quartile (25th percentile): <33%

Seed Stage:

- Top quartile (75th percentile): >47%
- Median (50th percentile): 27%
- Bottom quartile (25th percentile): <17%

Series A Stage:

- Top quartile (75th percentile): >56%
- Median (50th percentile): 43%
- Bottom quartile (25th percentile): <33%

Week 1 retention is often considered a leading indicator of product-market fit and the effectiveness of your onboarding experience. However, like other metrics, it should be viewed in context with your specific product, industry, and user behavior patterns.

Remember: These benchmarks should serve as rough guidelines rather than strict targets. Focus on understanding and improving your specific user journey rather than hitting industry averages that may not reflect your unique situation.

5. PRODUCT TOUR AND CHECKLIST COMPLETION RATES

Product tour and checklist completion rates measure how many users complete your guided onboarding experiences. These metrics help evaluate the effectiveness of your onboarding UX design and content.

According to Chameleon's 2025 Product Benchmark Report, here are the key completion rates for different types of product tours:

Overall Product Tour Completion:

- Average completion rate: 31%

By Trigger Type:

- Click-triggered tours (via checklist): 58.8%
- Auto-triggered tours (immediate display): 30%

By Number of Steps:

- Two-step tours: 37%
- Three-step tours: 35%
- Four-step tours: 41%
- Five-step tours: 22%
- Ten-step tours: 18%

The data shows a clear correlation between tour length and completion rates. Tours exceeding five steps see a sharp decline in completion, with more than half of users dropping off.

For onboarding checklists, around 23% of users explore and click within a checklist. Adding a Welcome Screen increases the checklist interaction by 27%.

An interesting finding is that users who engage with checklists tend to complete an average of five items per session, indicating a high level of commitment once they begin the process.

These benchmarks suggest shorter, user-initiated tours perform better than longer, automatically triggered ones. When designing your product tours and checklists, consider breaking longer flows into smaller, more digestible segments to maintain user engagement.